Nick Dixon, Susan Loxley and Paul Levy

REVISION PLUS

Edexcel

GCSE Science

ReviCompanion

Contents

Unit B1: Influences on Life

Unit C1: Chemistry in Our World

Unit P1: Universal Physics

Classification

There are millions of different species of living organisms on Earth and they are all **classified** (grouped) on the basis of their similarities and differences.

There are five groups called **kingdoms**: plantae, animalia, fungi, protoctista and prokaryotes (see table below). These kingdoms are divided into smaller, more specific groups. There are about 35 major groups called **phyla**. Phyla are then divided into smaller groups called **classes**. Classes are then divided into **orders**; orders are divided into **families**; families are divided into **genera**, and each genus is then split into individual **species**.

Species can be defined as groups of related, living organisms that have many features in common. Lions and tigers are different species.

A species can also be defined as organisms that can breed to produce fertile offspring. For example, lions can occasionally reproduce with tigers to make a hybrid liger, but ligers are not capable of reproducing themselves. There are limitations to this definition of species (see page 8). Some organisms do not always reproduce sexually and some hybrids are fertile.

The Binomial System

In the 18th century, Carl Linnaeus introduced the **binomial system** of classification.

This simple system gives each organism two names. Linnaeus used Latin for his classification system. It is so effective that scientists still use it today. The name for humans in the binomial system is *Homo sapiens*.

Kingdoms	
Plantae (Plants), e.g. Trees, Mosses	**Fungi, e.g. *Rhizopus*, *Agaricus***
<ul><li>Multicellular</li><li>Have cell walls</li><li>Have chlorophyll</li><li>Feed **autotrophically** (make their own food by photosynthesis)</li></ul> 	<ul><li>Multicellular or unicellular</li><li>Have cell walls</li><li>Do not have chlorophyll</li><li>Feed **saprophytically** (extracellular digestion of dead organic matter)</li></ul>
Animalia (Animals), e.g. Fish, Amphibians	**Protoctista, e.g. *Euglena*, *Paramecium***
<ul><li>Multicellular</li><li>Do not have cell walls</li><li>Do not have chlorophyll</li><li>Feed **heterotrophically** (unable to make their own food)</li></ul> 	<ul><li>Unicellular</li><li>Have a nucleus</li></ul> 
	Prokaryotes (bacteria), e.g. *Streptococcus*, *Staphylococcus*
	<ul><li>Unicellular</li><li>Do not have a nucleus</li></ul> 

Vertebrates and Invertebrates

The animal kingdom can be divided into smaller groups (phyla). Animals in the phylum **Chordata** have a supporting rod running the length of their body. In **vertebrates** this is called the backbone. **Invertebrates** do not have a backbone.

Vertebrates can be divided into five classes. Invertebrates are divided into a large number of phyla and classes. The vertebrate classes and some of the more important invertebrate phyla and classes are shown below.

Vertebrate Classes	Invertebrate Phyla and Classes
Fish	**Annelids (segmented worms)**
e.g. John Dory fish • Cold-blooded • Have gills • Lay eggs in water • Have wet scales	e.g. Common earthworm • Bodies divided into segments • Each segment has bristles
Amphibians	**Molluscs**
e.g. Frog • Cold-blooded • Adults have lungs (young have gills) • Lay eggs in water or very damp places • Have smooth, moist, permeable skin	e.g. Roman snail • Bodies unsegmented • Variable in shape • Some species of molluscs have shells
Reptiles	**Crustaceans**
e.g. Snake • Cold-blooded • Have lungs • Lay eggs • Have dry, scaly skin	e.g. Woodlouse • Body divided into three regions • Breathe through gills • External skeleton • Jointed legs
Birds	**Arachnids – spiders and scorpions**
e.g. Skylark • Warm-blooded • Have lungs • Lay eggs • Have feathers and a beak	e.g. Spider • Body divided into two regions • Four pairs of legs • External skeleton • Jointed legs
Mammals	**Insects**
e.g. Sheep • Warm-blooded • Have lungs • Give birth to live offspring • Produce milk • Have hair or fur on skin	e.g. Honey bee • Body divided into three regions • Three pairs of legs • Have wings • External skeleton • Jointed legs

Classification of Vertebrates

Scientists place vertebrates into groups based upon the way they:

- absorb oxygen (through lungs, gills and skin)
- reproduce (internal or external fertilisation and whether **oviparous** (animals that lay eggs) or **viviparous** (animals that develop their embryos internally))
- regulate their temperature (**homeotherms**, which maintain their body temperature, or **poikilotherms**, which cannot).

The Importance of Classification

There are currently well over one million named species of life on Earth. Some scientists think that there might be many times this total still to discover.

With this many species and such diversity of life, it is very important that organisms are carefully identified, named and then classified. Classification allows groups of similar organisms to be studied, making it easier to identify habitats or species that need **conservation**.

Conservation can keep **ecosystems** stable as environmental conditions change. These conditions include both **biotic factors** (living things) and **abiotic factors** (temperature, humidity, etc.).

Conservation can lead to greater **biodiversity** (the variety of different types of organism in a habitat or ecosystem) by:

- preventing species from becoming extinct
- maintaining variation between species
- preserving habitats.

Issues Surrounding Classification

Usually a species will fit into a group, but occasionally there are problems. Scientists find it difficult to classify all vertebrates based upon their anatomy and reproduction.

Scientists are not sure if viruses are alive, as they do not complete all of the seven life processes. As a result, classification of viruses is difficult – they do not fit into any of the five kingdoms.

Identification Keys

Keys can be used to identify many things, but are often used to find out the names of living organisms. They are a series of questions that have specific answers, which usually lead to another question. The questions continue until all of the answers are the names of organisms.

There are four species in the genus *Panthera*:

- tiger (*Panthera tigris*)
- lion (*Panthera leo*)
- jaguar (*Panthera onca*)
- leopard (*Panthera pardus*).

The key will allow you to identify one from another.

1. Does it have a mane? If yes, this is *Panthera leo*. If not, move onto the next question.
2. Does it have stripes? If yes, this is *Panthera tigris*. If not, move onto the next question.
3. Does it have large spots with smaller spots in the middle? If yes, this is *Panthera onca*. If not, move onto next question.
4. Does it only have small spots with no spots in the middle? If yes, this is *Panthera pardus*.

Adaptations

Adaptations are special features or types of behaviour that make a living organism well-suited to its environment.

Adaptations develop as the result of **evolution**. They increase an organism's chance of staying alive.

Adaptations are **biological solutions** to the challenge of survival.

Examples of Adaptations

A Cold Terrestrial Climate – Polar Bear

- Small surface area to volume ratio to reduce heat loss.
- Large amount of insulating fat beneath skin.
- Large feet to spread its weight on the ice.
- Powerful legs so it can swim to catch food.
- White coat so it is camouflaged.

A Hot Terrestrial Climate – Camel

- Large surface area to volume ratio to increase heat loss.
- Body fat stored in hump (almost no fat beneath skin).
- Sandy coat so it is camouflaged.
- Loses very little water through urine and sweating.
- Can drink up to 20 gallons of water at once.

An Aquatic Environment – Fish

- Streamlined shape to travel quickly through water.
- Gills obtain dissolved oxygen from the water.
- Gills have a large surface area to increase the amount of oxygen that can be absorbed.

A Hot Terrestrial Climate – Cactus

- Thick, waxy surface to reduce water loss.
- Spines to protect it from predators who would eat it for its water.
- Stomata only open at night to reduce water loss.
- Some have shallow-spreading roots to absorb surface water; others have deep-spreading roots to tap into underground water supplies.

An Aquatic Environment – Water Lily

- Flexible stem so that it can bend in the water's current.
- Underwater leaves are streamlined.
- Leaves often grow on the surface of the water to maximise photosynthesis.

Extreme Habitats

Different living things are adapted to live in all parts of the world. Some habitats on Earth are described as extreme. **Extreme habitats** have conditions outside those in which normal organisms live. Organisms living in extreme habitats need to have very special adaptations to be able to survive. Some examples are given below:

Hydrothermal Vents

Found at the bottom of the oceans, hydrothermal vents discharge very hot water (from 60°C to over 400°C). As it is completely dark, there are no photosynthesising plants. However, there are many different types of organisms that are adapted to survive here by:
- being able to cope with the high pressure and temperature
- having very highly developed senses other than sight.

The Antarctic

The temperatures are very low in the Antarctic; they can fall to below -50°C. Not many animals can survive in such difficult conditions. However, penguins are adapted to survive because they:
- have a compact shape, meaning that relatively little heat is lost through their surface (small surface area to volume ratio)
- have a thick layer of insulating fat under their skin
- have very tightly packed, waterproof feathers for insulation
- often huddle in large, tightly packed groups (rookeries); they constantly move into the centre so that the ones on the outside are not there for long and, therefore, never get too cold.

Variation

Variation is the differences between individuals of the same species. If we look at the faces of our friends we can see variation in skin and eye colour, and the shape and length of our noses. Variations within species can make classification difficult.

Complications in Classification

An example where classification becomes complicated is **hybridisation** in ducks. A hybrid duck has two different species as parents and is fertile, so could be considered a new species (see page 4).

Ring species also make classification difficult. These are overlapping populations of two closely-related but different species that can only interbreed in the overlapping region. Populations of salamanders on the South West Coast of the USA are famous examples of ring species.

Speciation

Speciation is the process of evolution by which new species form. This often results from sections of populations being isolated by geographical features. A famous example is the diversity now seen in the three-spined stickleback. After the last Ice Age this marine fish underwent speciation into new freshwater colonies in geographically isolated lakes and streams.

Inheritance v Environment

Variation is due to two factors:
1. **Genetic causes** (**inheritance**): variation due to the genes inherited by an individual after sexual reproduction or genetic **mutation**.
2. **Environmental causes**: variation due to the conditions in which an individual has developed. Characteristics obtained in this way are called **acquired**.

There has been a long and lively debate over the relative importance of genetic and environmental factors in determining certain characteristics. Most people now agree that variation is due to a combination of genetic and environmental causes. The table below illustrates this.

Feature	Genetic Factors	Environmental Factors
Sporting ability	<ul><li>A person's natural sporting ability and co-ordination.</li><li>Natural physique and body structure.</li></ul>	<ul><li>Good coaching.</li><li>Positive support.</li><li>Excellent facilities and opportunities to practise.</li></ul>
Intelligence	<ul><li>The structure of the brain and its nerve connections.</li></ul>	<ul><li>Support from home.</li><li>Quality of education.</li><li>Life experience.</li></ul>

Investigating Continuous and Discontinuous Variation

Variation in any species can be grouped into two types:
- continuous
- discontinuous

Here we will look at these types in humans.

Discontinuous Variation
Discontinuous variation fits into a small number of clearly defined groups. A common example is eye colour. There are only three groups of eye colour and they do not have a range of values, i.e. you either have blue, brown or green eyes. Another example is blood type. Results from investigations into discontinuous variation are shown in a **bar graph**.

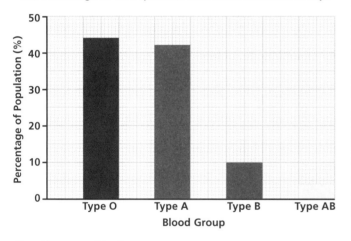

The Percentage of UK Population with Different Blood Groups

Continuous Variation
Continuous variation does not fit into groups so easily and values have a range. Common examples include height, weight and skin colour. Results from investigations into continuous variation are shown in a **line graph**.

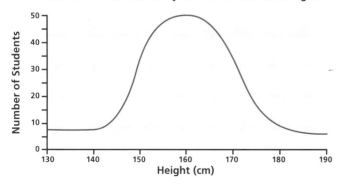

The Number of Students In My Year With Different Heights

If the sample size is large enough a normal distribution (as in the previous graph) will be seen.

Evolution

The theory of evolution states that all living organisms alive today (and those that are now **extinct**) developed from simple life forms.

When species evolve, they become better adapted to their environment. Species that are less well-adapted may become extinct.

Charles Darwin (1809–1882) deduced that all organisms are involved in a struggle for survival and that only the best-adapted survive. Organisms that survive are more likely to reproduce and pass their (well-adapted) genes on to their offspring. The rest die out. So, very gradually, a species will change. This formed the basis of Darwin's **theory of evolution by natural selection** (see page 10).

At first Darwin had great difficulty in having his theory accepted because it is difficult to prove, and many scientists did not (and some still do not) accept the theory. Most rejected it because it contradicted the Bible. (In those days the Church had great influence over what people thought.)

Recently many scientists have used DNA research to support Darwin's theory. The emergence of antibiotic-resistant bacteria (see page 23) also supports it.

Scientists all over the world now work to support, disprove or improve scientific evidence by:
- writing their findings in scientific journals
- peer-reviewing each other's research
- presenting their work at scientific conferences.

Natural Selection

Evolution can be explained by **natural selection**. There are six steps:

1. **variation**: most populations of organisms have individuals that vary slightly
2. **over-production**: most organisms produce more young than will survive to adulthood
3. **struggle for existence**: population sizes are generally stable so there must be competition between organisms
4. **survival**: those with advantageous characteristics are more likely to survive (survival of the fittest)
5. **inheritance**: better adapted organisms are more likely to reproduce and pass on their characteristics
6. **gradual change**: over time more individuals in a population will possess the advantageous characteristics (and poorly adapted characteristics may be lost).

Organisms that are not well-adapted to their environment could become extinct.

Chromosomes and Genes

In the **nucleus** of a normal human body cell there are two sets of **chromosomes**. One set comes from the mother and the other from the father. Each set consists of 23 individual chromosomes. So each cell has 46 chromosomes (23 pairs) in total. Other animals and plants have different numbers (e.g. dolphins have 44 and cabbages have 18).

Chromosomes are divided into sections called **genes**. It is our genes that control what characteristics we have. Chromosomes (and therefore genes) are made from a chemical called **DNA**.

Alleles

A gene is the part of a chromosome that controls the development of a characteristic, either individually or in combination with other associated genes. For each gene, there can be different versions called **alleles**.

As we already know, normal body cells (not gametes) contain pairs of chromosomes. This means that genes also exist in pairs; one on each chromosome. So, for each gene an individual might have two matching alleles or two different alleles.

A **dominant allele** is an allele that controls the development of a characteristic, even when it is only present on one of the chromosomes in a pair.

A **recessive allele** only controls the development of a characteristic if it is present on both of the chromosomes in a pair.

Genetic Variation

For each pair of genes there are several possible combinations of alleles, which produce different outcomes. For example:

* The gene that controls eye colour has two alleles: one for brown eyes and one for blue eyes.
* The gene that controls earlobe type has two alleles: one for attached earlobes and one for unattached (free) earlobes.

When both alleles are dominant, the individual is described as being **homozygous dominant** for that gene or outcome.

When both alleles are recessive, the individual is described as being **homozygous recessive** for that gene or outcome.

When there is one dominant allele and one recessive allele, the individual is described as being **heterozygous** for that gene or outcome.

So, the possible combinations are as follows:

	Homozygous Dominant	Heterozygous	Homozygous Recessive
Earlobes	EE (free lobes)	Ee (free lobes)	ee (attached lobes)
Eye colour	BB (brown eyes)	Bb (brown eyes)	bb (blue eyes)

Family Pedigrees

The presence or absence of specific alleles from one generation to the next can be shown in family pedigree charts. These 'family tree' diagrams are most often made for humans, show dogs and racehorses.

Monohybrid Inheritance

When a characteristic is determined by just one pair of alleles then simple genetic crosses can be performed to investigate the mechanism of inheritance. These simple crosses are examples of **monohybrid inheritance**.

These crosses can be shown using two types of genetic diagram: a grid (**Punnett square**) or a genetic cross diagram.

In all genetic diagrams we use letters to simplify words, e.g. for earlobes, combinations of E and e are used. These are called **genotypes**. The physical characteristics displayed by the individual are called **phenotypes**, e.g. attached earlobes. In genotypes, we use capital letters for dominant alleles and lower case letters for recessive alleles.

Remember to identify clearly the alleles of the parents, separate each pair of alleles into individual gametes and show all possible combinations, i.e. each gamete from the father should be paired with each gamete from the mother in turn.

In diagrams **1** and **2**, one parent has two dominant alleles, so each offspring will inherit the dominant feature.

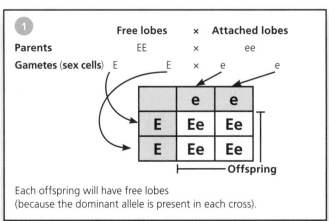

Each offspring will have free lobes
(because the dominant allele is present in each cross).

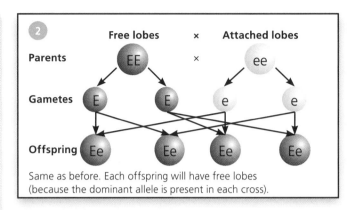

Same as before. Each offspring will have free lobes
(because the dominant allele is present in each cross).

In diagram **3** each parent has one dominant allele and one recessive allele, so each offspring will have a 3 in 4 (75%) chance of inheriting the dominant feature. In diagram **4** one parent has one recessive allele and the other parent has two recessive alleles, so each offspring will have a 1 in 2 (50%) chance of inheriting the dominant feature.

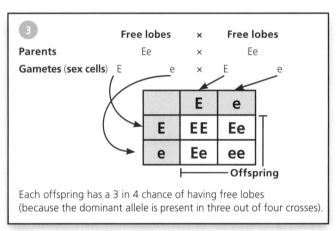

Each offspring has a 3 in 4 chance of having free lobes
(because the dominant allele is present in three out of four crosses).

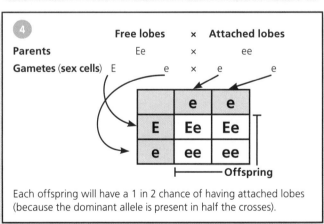

Each offspring will have a 1 in 2 chance of having attached lobes
(because the dominant allele is present in half the crosses).

The possible outcomes for one pair of alleles are limited, but there are numerous genes on the chromosomes in human body cells. This is why individuals of the same species (even brothers and sisters) show variation.

Probabilities from Monohybrid Crosses

Probabilities from monohybrid crosses usually involve 'wordy' descriptions, which you have to translate into crosses.

Example 1
Draw genetic diagrams to predict the probable genotypic ratios produced when two heterozygous brown-eyed people mate.
Brown eyes are dominant to blue eyes.

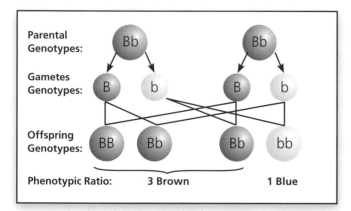

Parental Genotypes:	Bb		Bb	
Gametes Genotypes:	B	b	B	b
Offspring Genotypes:	BB	Bb	Bb	bb
Phenotypic Ratio:	3 Brown			1 Blue

The genetic cross diagram reveals a 3 : 1 ratio of brown eyes to blue eyes.

Example 2
In mice, white fur is dominant to grey fur. What ratio of phenotypes would you expect to be produced from a cross between a heterozygous individual and one with grey fur?

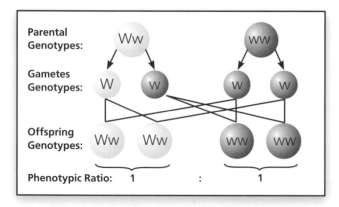

Parental Genotypes:	Ww		ww	
Gametes Genotypes:	W	w	w	w
Offspring Genotypes:	Ww	Ww	ww	ww
Phenotypic Ratio:	1	:		1

There is a 1 : 1 ratio of white fur to grey fur.

Symptoms of Genetic Disorders

You need to know the symptoms of the following genetic disorders.

Cystic Fibrosis
Cystic fibrosis is an inherited disorder that affects cell membranes.

Characteristics of the disorder include:
- being unable to digest food properly
- airways becoming clogged with excess, sticky mucus
- more likely to suffer from chest infections, including pneumonia
- shorter life expectancy – sufferers often die aged between 40 and 50.

There is no cure for cystic fibrosis, but the symptoms can be treated using enzyme tablets, chest physiotherapy, and antibiotics for chest infections.

Sickle Cell Disease
Sickle cell disease is an inherited disorder that affects red blood cells. Instead of being their normal biconcave shape, they look more like sickles. They are no longer able to carry oxygen and can get stuck in capillaries. This can stop blood reaching tissues, which can cause great pain and lead to tissue damage. It is very serious if this happens in the lungs. Symptoms of sickle cell disease include shortness of breath and dizziness.

Genetics of Genetic Disorders

Inherited or **genetic disorders** like sickle cell disease and cystic fibrosis are caused by 'faulty' alleles that can come from one or both parents.

Sickle cell disease occurs when a person inherits a 'faulty' recessive allele from each parent. Individuals who have only one faulty allele will not have the disorder but the faulty allele could be passed on to their offspring. These individuals are **carriers**.

Cystic fibrosis is inherited in the same way as sickle cell disease. It is unusual for people with this condition to have children, so most instances arise from two parents who are carriers.

B1 Topic 2: Responses to a Changing Environment

This topic looks at:

- how humans detect and respond to changes in their internal and external environments
- the role of hormones and the nervous system in detecting and responding to these changes
- the role of hormones in plant responses to stimuli

Homeostasis

The human body is made up from millions of cells. To function properly cells need certain conditions. Keeping these conditions the same is called **homeostasis**. Scientists call this maintaining a stable internal environment.

Three very important examples of homeostasis are:

- keeping the correct levels of **water** in the body
- keeping a constant **temperature**
- keeping the correct amount of **glucose** in the blood (page 17).

Regulating Water (Osmoregulation)

When we have not consumed enough water we urinate less and our urine is more concentrated (it is darker yellow in colour). When we are properly hydrated we urinate more often and our urine is more diluted. It is clear or lighter yellow.

Regulating Body Temperature (Thermoregulation) and the Skin

The skin is the largest organ in the body and plays a very important role in regulating body temperature.

Since enzymes work best at 37°C (in humans), it is essential that the body remains very close to this temperature. Monitoring and control is done by the **hypothalamus** in the brain, which has receptors that are sensitive to the temperature of the blood flowing through it.

There are also temperature receptors in the skin, which provide information about skin temperature.

Core temperature too high

- Blood vessels in skin dilate (become wider) causing greater heat loss from the skin.
- Sweat glands release sweat (mainly water and salts) which evaporates, removing heat from the skin.
- Erector muscles cause hairs to lie flat and not trap heat.
- In hot and dry conditions, sebaceous glands produce oily sebum to encourage sweat to spread effectively.

Hypothalamus

Core temperature too low

- Blood vessels in skin constrict (become narrower) reducing heat loss from the skin.
- Muscles start to 'shiver', causing heat energy to be released via respiration in cells.
- Erector muscles cause hairs to stand up, trapping heat.
- In cool and wet conditions, sebaceous glands produce oily sebum that waterproofs the skin, helping excess water to run off.

HT Vasodilation and Vasoconstriction

The body can control the amount of blood that is near the surface of the skin. When the body is too warm, more blood flows near the surface through superficial capillaries. This allows more heat to be lost, which cools the body down. This is called **vasodilation**.

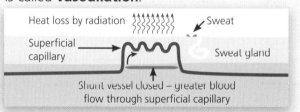

If the body is too cold, the blood flow through the superficial capillaries is restricted near the skin surface. Less heat is lost, which helps keep the body warmer. This is called **vasoconstriction**.

This is an example of **negative feedback**.

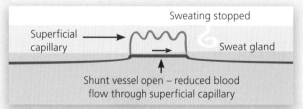

The Nervous System

The **nervous system** consists of the **brain**, the **spinal cord**, **paired nerves** and **receptors**. The brain and the spinal cord are referred to together as the **central nervous system** (CNS). The rest of the nerves in the body are collectively called the **peripheral nervous system**.

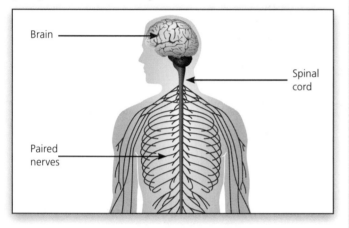

The nervous system allows organisms to react to their surroundings and coordinate their behaviour.

Humans have five senses. Receptors in the sense organs detect internal and external changes, allowing the body to respond to these stimuli:

- sight (eyes)
- hearing (ears)
- taste (taste buds on the tongue)
- smell (chemical receptors in the nose)
- touch (receptors in the skin).

N.B. Balance (ears) may also be referred to as a sense.

The Brain

The **brain** coordinates most of the body's actions.

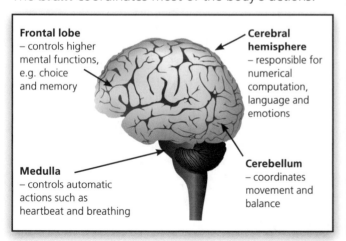

Frontal lobe – controls higher mental functions, e.g. choice and memory

Cerebral hemisphere – responsible for numerical computation, language and emotions

Medulla – controls automatic actions such as heartbeat and breathing

Cerebellum – coordinates movement and balance

Neurones

Neurones are specially adapted cells that carry electrical signals called **impulses**. Neurones are elongated (stretched out) to make connections between parts of the body. The elongated parts of neurones are covered by an insulating layer called the **myelin sheath**. This increases the speed of electrical impulses passing along them.

The elongated parts of neurones that carry impulses away from the cell body are called **axons**; those that carry impulses towards the cell body are called **dendrons**. Both axons and dendrons usually have branched endings, which allow a single neurone to act on many muscle fibres or connect with many other neurones.

1. **Sensory neurones:** these take nerve impulses from stimulated receptors in sense organs to the central nervous system.

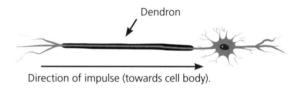

Dendron

Direction of impulse (towards cell body).

2. **Relay neurones:** in the central nervous system these pass impulses on from sensory neurones to motor neurones.

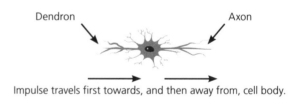

Dendron   Axon

Impulse travels first towards, and then away from, cell body.

3. **Motor neurones:** these take impulses from the central nervous system to the muscles or glands.

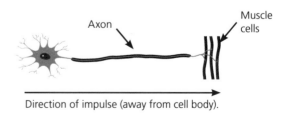

Axon   Muscle cells

Direction of impulse (away from cell body).

Synapse

Neurones pass information into and out of the central nervous system. Neurones do not touch each other. There is a very small gap between them called a **synapse**.

1. When an electrical impulse reaches this gap via neurone A, a chemical transmitter (**neurotransmitter**) is released that activates receptors on neurone B.

2. This causes an electrical impulse to be generated in neurone B.

3. The chemical transmitter is then destroyed or removed.

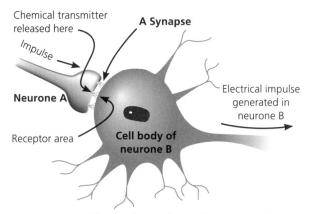

So, an impulse will go along the following pathway:

Sense organ
Receptors detect a change either inside or outside the body. This change is a stimulus.

Sensory neurone
Conducts the impulse from the sense organ towards the C.N.S.

Synapse
The gap between the sensory and relay neurones.

Relay neurone
Passes the impulse on to a motor neurone.

Synapse
The gap between the relay neurone and the motor neurone.

Motor neurone
Passes the impulse on to the muscle (or gland).

Muscle
The muscle will respond by contracting, which results in a movement. Scientists call muscles (and glands) **effectors**.

Investigating Skin Sensitivity

There are many receptors on the surface of the skin. Amongst other things they can detect touch, temperature, pressure and pain. The receptors in some parts of the skin are much closer together than in other parts.

The distance between receptors can be estimated by gently touching two points (perhaps an unwound paperclip) onto the surface of the skin. Start reasonably far apart and move the two points closer together. When you can only feel one point (not two) you have discovered the distance that the receptors are apart.

Lips and fingertips are especially sensitive to touch, which means the receptors are very close together.

Voluntary and Reflex Responses

Voluntary Responses (Actions)

These are actions over which we have complete control i.e. we consciously decide to act. For example, speaking, walking or picking something up.

Stimulus	Receptors	Coordinator	Effectors	Response
Freshly baked cake	Sight and smell receptors	**Sensory Neurones** ▼ **CNS** ▼ **Motor Neurones** ▶	Muscles in hand	Move hand to pick up cake

Involuntary/Reflex Responses (Actions)

These are responses over which we have no control i.e. they happen automatically. For example, blinking or moving part of your body away from pain.

Stimulus	Receptors	Coordinator	Effectors	Response
Drawing pin	Pain receptor (in finger)	**Sensory Neurones** ▼ **Relay Neurone in Spinal Cord** ▼ **Motor Neurones** ▶	Muscles in hand	Withdraw hand

Reflex Arcs

Sometimes conscious action would be too slow to prevent harm to the body. Reflex action speeds up the response time by missing out the brain. The spinal cord acts as the coordinator and passes impulses directly from a sensory neurone to a motor neurone via a relay neurone, which bypasses the brain. This is called a **reflex arc** (see diagram below).

1. A receptor is stimulated by the sharp drawing pin (stimulus)…
2. …causing impulses to pass along a sensory neurone into the spinal cord.
3. The sensory neurone synapses with a relay neurone.
4. The relay neurone synapses with a motor neurone, bypassing the brain and sending impulses down the motor neurone…
5. …to the muscles (effectors), causing them to contract in response to the pain from the sharp drawing pin.

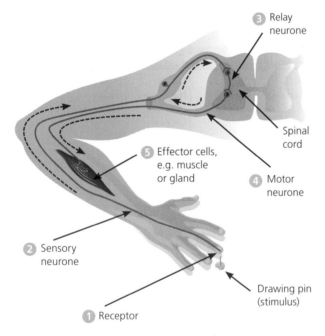

In the diagram, dashed black arrows show the direction of the impulses.

Investigating Human Responses to External Stimuli

To study responses to voluntary stimuli, you can carry out the ruler-drop investigation (see page 21). To study responses to involuntary stimuli you can try the following:

- **Iris reflex** – carefully shine a light into one of your partner's eyes. Observe how the pupil becomes smaller.
- **Temperature control** – either cool or warm your partner's skin on their arm and observe the response (see page 13).

Hormones and Coordination

Hormones are chemical messages, produced by **endocrine glands**. Hormones coordinate and control the way in which parts of the body (target organs or target cells) function. Hormones are transported to their **target organs** or **target cells** through the bloodstream. For example:

- The adrenal glands (above the kidneys) produce **adrenaline**.
- The pituitary gland (in the brain) produces a **growth hormone**, **FSH** (follicle-stimulating hormone) and **LH** (luteinising hormone).
- In males the testes produce **testosterone**.
- In females the ovaries produce **oestrogen** and **progesterone**.
- The pancreas produces **insulin**, e.g.

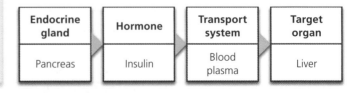

Endocrine gland	Hormone	Transport system	Target organ
Pancreas	Insulin	Blood plasma	Liver

Control of Blood Glucose

The body needs controlled quantities of glucose for respiration.

Different parts of the body work together to monitor and control blood glucose levels.

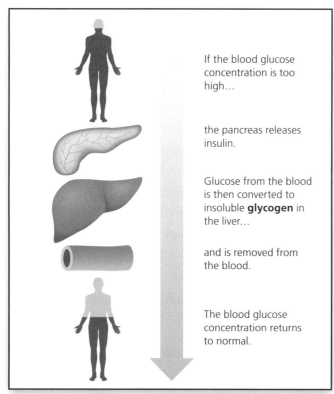

If the blood glucose concentration is too high…

the pancreas releases insulin.

Glucose from the blood is then converted to insoluble **glycogen** in the liver…

and is removed from the blood.

The blood glucose concentration returns to normal.

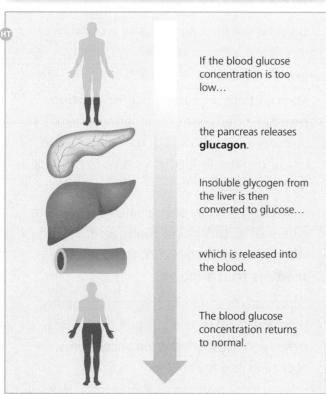

If the blood glucose concentration is too low…

the pancreas releases **glucagon**.

Insoluble glycogen from the liver is then converted to glucose…

which is released into the blood.

The blood glucose concentration returns to normal.

Diabetes

Diabetes is a condition where the amount of glucose in the blood is too high because the cells cannot absorb it and use it properly. There are two types of diabetes. **Type 1 diabetics** are unable to produce insulin. **Type 2 diabetics** become resistant to the insulin they produce.

The main symptoms of diabetes are:
- urinating more often than normal
- becoming more thirsty
- increased tiredness
- weight loss
- blurred vision.

Treatment of Diabetes

Diabetes cannot yet be cured. It is important for both Type 1 and Type 2 diabetics to lead **healthy lifestyles** by:
- eating three meals a day
- including some carbohydrate but reducing fat and sugar in meals
- being physically active.

Doing this may be enough to control Type 2 diabetes, at least in its early stages. However, people who have Type 1 diabetes will also need to **inject insulin** into the fat layer beneath their skin (subcutaneous) to control their blood sugar concentration. Before injecting insulin, a person with Type 1 diabetes will test the amount of sugar in their blood. If they have eaten food containing a lot of sugar then a bigger dose of insulin is required. If they are going to be very active and use up a lot of blood glucose then a smaller dose of insulin is needed.

Obesity can occur when a person eats too much fatty food and/or does not exercise enough. They become very overweight. Scientists now think that obesity can lead to Type 2 diabetes. Having a healthy diet and being physically active are very important in not becoming obese and so possibly diabetic. Doctors use **BMI** (**Body Mass Index**) calculations, based on height and weight, to monitor obesity.

Plant Hormones

Plants are sensitive to light and gravity. Shoots grow towards the light (**positively phototropic**) and away from gravity. Roots grow in the direction of gravity (**positively geotropic** or **gravitropic**). These responses are controlled by hormones that coordinate and control growth.

The two types of hormones you need to know about are called **auxins** and **gibberellins**. Auxins are plant hormones that specifically promote cell **elongation**. Gibberellins promote cell elongation but they also encourage flowering, breaking seed dormancy and can be used to force seedless fruit to develop (see below).

Gravity

Hormones in solution travel down towards the lower side of shoots and roots:

- In the shoots, auxins increase growth in the lower region, which makes the shoot bend upwards.
- In the roots, the hormones slow down growth in the lower region, which makes the root bend downwards.

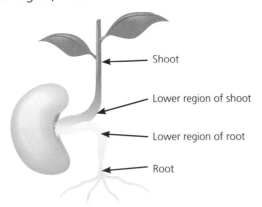

Shoot

Lower region of shoot

Lower region of root

Root

Light

In shoots, light causes auxins to accumulate on the shaded part of the stem, which causes growth on that side and means that the plant grows towards the light.

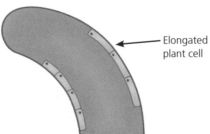

Light source

Elongated plant cell

Investigating Phototropism in Cress Seedlings

Cress seedlings are positively phototropic. Investigate this by completing the following three experiments, which involve growing cress seeds on damp cotton wool.

1. Leave the first dish in a sunny area, like a windowsill. The seedlings should develop normally. They should be tall, dark green and growing reasonably upright.
2. Place the second dish in the same area but completely cover it with a larger dish to block out all the light. The seedlings should be longer, thinner, paler green and growing in different directions to try and find the light.
3. Place the third dish in the same area. Cover it with a similar large dish to the second experiment, but with a small hole to allow light through. The seedlings should again be longer, thinner and paler green, but this time growing towards the hole that is letting light in.

HT Commercial Uses of Plant Hormones

Gardeners often apply **selective weedkillers** to their lawns, which contain a hormone that disrupts the growth of all plants other than grass.

Stem cuttings are often placed in **rooting powder**, which contains a hormone to promote the formation of roots. This allows lots of genetically identical plants to be made from one plant.

Fruit growers spray unpollinated flowers with gibberellins. This makes plants produce fruit without fertilisation occurring. As a result, **seedless fruit** is grown.

Supermarkets and fruit importers use another hormone called ethene to **ripen fruit**. Fruit is picked and transported when unripe and is ripened at its destination.

This topic looks at:
- how the body is affected by drugs and pathogens
- how scientists have helped in the development of antibiotics and antiseptics
- the interdependence of organisms on Earth
- how man-made chemicals can pollute the environment
- the carbon and nitrogen cycles

How Drugs Can Affect You

Drugs are chemical substances that affect the central nervous system, causing changes in psychological behaviour and possibly addiction. Some drugs are obtained from living things (often plants) whilst others are synthetic (man-made). Many drugs are **medicines** used to cure illnesses or ease symptoms. Examples are painkillers and antibiotics.

Paracetamol

Paracetamol is a commonly used **painkiller** and **anti-inflammatory**, which also helps to lower body temperature. Great care must be taken not to exceed the recommended dosage as overdosing can lead to liver failure and death.

Cannabis and Opiates

Cannabis comes from cannabis plants. **Opiates** are drugs that come from poppy plants. Both cannabis and opiates are able to relieve pain, especially in the terminally sick. However, they are both illegal drugs. In recent years there has been a lot of scientific research into how effective cannabis is at controlling pain. Many doctors and scientists believe there is enough evidence to support the use of cannabis in strictly controlled circumstances. Others feel there still needs to be more research conducted into the benefits and the long-term effects.

Drug	Effect on Nerve Transmission and Reaction Time	Effect on Activities	Abnormal Behaviour Caused
Stimulant (e.g. caffeine and nicotine)	They speed up the transmission of a message across a synapse (see p.15).	Reactions to things going on around you could become faster.	They can keep you awake leading to physical exhaustion, and can make you highly strung.
Depressant (e.g. alcohol and barbiturates)	They slow down the transmission of a message across a synapse.	They can make you drowsy so you must not drive or do any other activity requiring full concentration.	They are addictive. They can make you irritable, aggressive, confused and easily upset.
Painkillers (e.g. paracetamol and morphine)	They prevent the transmission of a message across a synapse.	They can make you drowsy and/or cause blurred vision so you must not drive or do any activity requiring full concentration.	Side effects include feeling dizzy, or feeling itchy all over.
Hallucinogens (e.g. LSD)	Can block pathways to the sensory parts of the brain resulting in hallucinations.	They distort sense perception so you must not drive or do any activity requiring full concentration.	They cause hallucinations which can by make you scared, confused and easily upset.

Solvents, Alcohol and Tobacco

Solvents, alcohol and tobacco can have major physical and mental effects on the human body.

Substance	Effects
Solvents are depressants that give off different kinds of vapours.	**Physical Effect** (when inhaled) • Can cause permanent damage to the lungs, liver, brain and kidneys. **Mental Effects** • Increases reaction times by slowing down transmission of a message across synapses (see page 15). • Cause hallucinations. • Alter behaviour and personality.
Alcohol is a depressant that contains the chemical ethanol.	**Physical Effects** • Short-term use can cause blurred vision. • Long-term use can lead to brain damage and liver damage (cirrhosis), because the liver removes alcohol from the body. Excess use can lead to unconsciousness, brain damage, coma and death. **Mental Effects** • Increases reaction times by slowing down transmission of a message across synapses. • Can cause depression. • Can lead to loss of inhibitions and self-control.

Tobacco contains tar and nicotine, and produces carbon monoxide when smoked.	**Physical Effects**
	• **Carbon monoxide** in smoke is absorbed by haemoglobin in red blood cells more easily than oxygen.
	• Long term use can lead to:
	  – **emphysema:** alveoli walls break down
	  – build-up of mucus because cilia stop moving
	  – bronchitis and other chest infections
	  – damage to the circulatory system: damaged blood vessels, which can lead to heart attacks, strokes, arterial and heart disease and even amputations of legs
	  – **cancer** caused by **tar**
	  – increased heart rate and blood pressure caused by nicotine narrowing blood vessels.
	Mental Effect
	• **Nicotine** is addictive.

Cancer

The tar in cigarette smoke contains **carcinogens** (chemicals that cause cancer). These can cause cells to mutate and divide uncontrollably, which can form **tumours**. The tar can cause cancer of the lungs, throat and mouth.

Smoking and Poor Health

In recent years, many studies have shown a strong **correlation** (statistical relationship) between smoking and poor health. The correlation is particularly strong between smoking and lung cancer.

Investigating Reaction Times

Reaction times can be investigated in a number of ways. Perhaps the easiest way is to work in pairs dropping and catching rulers.

1. One person should sit at a table with their arm resting on it and only their open hand over the edge.
2. The second person should hold the bottom of the ruler in line with the top of the first person's thumb and finger. They should drop the ruler without giving any warning.
3. The first person should catch it between their thumb and first finger as quickly as possible.

This should be repeated a number of times and an average calculated. The average distance the ruler travels before it is caught can be converted into a reaction time using a table.

Transplants

A **transplant** is when an organ or tissue is moved from one patient to another or from one part of a patient to another. **Organs** that can be transplanted are hearts, kidneys, livers, lungs, pancreases and intestines. **Tissues** that can be transplanted are bones, tendons, corneas, skin, heart valves and veins. Donors for transplants can be living or very recently dead.

Ethical Issues of Organ Transplants

Organ donation is a controversial issue and some people have very strong feelings about it. The following are common examples of **ethical** issues:

- Voluntary or mandatory donation when we die (we currently volunteer to do this, commonly by carrying an organ donor card).
- Should people who may have had some control over their condition be allowed to have transplants (e.g. liver transplants for alcoholics and heart transplants for clinically obese people)?
- Theft of, and then the illegal trade of, organs (trafficking).

Pathogens and Disease

A **pathogen** is a microorganism that causes a disease. There are three main types: bacteria, fungi and viruses.

Bacteria

e.g. tuberculosis (TB), salmonella.
Treated by antibiotics.

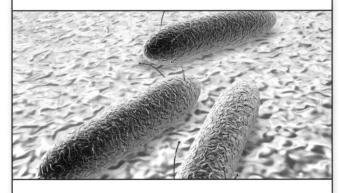

Fungi

e.g. athlete's foot, ringworm.
Treated by anti-fungal medicine and antibiotics.

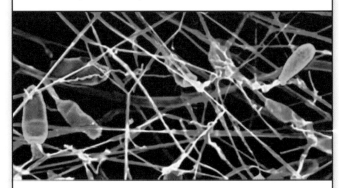

Viruses

e.g. Flu, chicken pox, HIV.
Very difficult to treat.

Transmission of Pathogens

Pathogens can be **transmitted** (passed from one organism to another) in many ways. The most common types of transmission are **indirect**.

Cholera is an example of a disease caused by a bacterium *Vibrio cholerae* that is transmitted in **dirty drinking water.** Cholera mainly affects the small intestine and its symptoms are diarrhoea and vomiting.

Salmonella is an example of a disease caused by lots of different species of *Salmonella* bacteria that are transmitted in **food**. It affects the intestine and its symptoms are diarrhoea and vomiting.

Many pathogens are transmitted though the air, commonly after coughing or sneezing. They are said to be **airborne**. Influenza is an example of an infection caused by a virus that is airborne. The main symptoms of influenza are chills or fever, sore throats, muscle pains and headaches.

Some pathogens are transmitted by another organism (the **vector** for the pathogen). For example, malaria is caused by infection from a malarial protozoan. If someone who has malaria is bitten by an *Anopheles* mosquito, the disease can be spread to all the other people the mosquito bites. Another example is the spread of the dysentery bacterium by the housefly.

Other pathogens are transmitted by **direct contact**. An example of a fungal disease that is transmitted by direct contact is the athlete's foot fungus. This causes itching and flaking of the skin, creating open wounds.

HIV is an example of a viral disease that is transmitted from one person to another by **body fluids**. Body fluids are sometimes exchanged during sexual intercourse or by reusing needles (often by drug addicts). HIV is the virus that causes AIDS. Sufferers of AIDS have severely weakened immune systems and so other infections, which healthy people fight off, can kill them more easily.

Plants' Defences and Medicines

Some plants defend themselves from attack from pathogens and herbivores by making **protective chemicals**. The study of some of these has led to the discovery of new drugs. Most of these medicines are now made synthetically, but they originated in plants, some a very long time ago.

Drug and its Source	Use
Aspirin Found in the bark and leaves of willow plants (a compound called salicin).	**Pain relief**, e.g. headaches, muscle and joint pain. It works by inhibiting the production of certain chemicals that transmit 'pain information' to the brain.
Taxol Found in the bark of the Pacific yew tree.	Treatment of some forms of **cancer**, e.g. breast cancer and ovarian cancer. It works by disrupting mitosis so the tumour cells cannot divide.

Animals' Defences

Animals (including humans) defend themselves from attack by pathogens with physical and chemical barriers.

Physical barriers include:
- **skin** – keeps pathogens outside the body
- **mucus** – sticky mucus in the airways of the respiratory system traps microorganisms
- **cilia** – these tiny hairs, which line the airways, beat rhythmically to remove the mucus.

Chemical barriers include:
- **lysozyme** – enzymes found in tears that destroy bacteria
- **hydrochloric acid** in the stomach – this strong acid destroys bacteria that have been swallowed.

Antiseptics

Antiseptics are substances that are applied to surfaces, commonly broken skin, to prevent infection by microorganisms. Frequently used antiseptics include alcohol, iodine and sodium chloride (salt).

Antibiotics

Antibiotics are medicines that kill or inhibit the growth of **bacteria** or **fungi**. They do not treat diseases caused by viruses, such as the common cold.

The first antibiotic, **penicillin**, was discovered by the Nobel Prize winning scientist **Sir Alexander Fleming** in 1928. He was investigating the bacterium *Staphylococcus* and, after returning from holiday, discovered that the fungus *Penicillium notatum* had contaminated his Petri dishes. Where the fungus grew, the bacteria did not. He studied the fungus further and isolated the naturally occurring penicillin antibiotic.

Penicillin is an example of one of many **antibacterial** medicines that specifically kill bacteria. Other antibiotics kill fungi and so are called **antifungals**. Other substances are antibacterial and antifungal. A common example of this type of substance is tea tree oil, which occurs naturally in the leaves of an Australian tree.

Investigating the Effects of Antibiotics and Antiseptics

The effects of antibiotics and antiseptics can be investigated in the following way.

1. Take an agar plate and spread a bacterial culture evenly across it using a sterile inoculating loop.
2. Incubate the agar plate until the bacteria cover the surface.
3. Divide your agar plate into equal sections.
4. Label your sections with the types of antibiotic or antiseptic solutions that you are investigating.
5. Apply some of the antibiotic or antiseptic solutions to small disks of filter paper and place them in the middle of the appropriate sections of your agar plate.

6. Several days later, measure the size of the clearing around the disk where the bacteria have been killed. The larger the clearing, the more effective the antibiotic or antiseptic is.

MRSA and the Misuse of Antibiotics

In recent years, so-called '**superbugs**' have evolved. These are bacteria that have become resistant to some of our antibiotic medicines. An example of a superbug is **MRSA**, which stands for methicillin-resistant *Staphylococcus aureus*. This bacterial infection grows in the nostrils and respiratory tract. It causes small red spots, which can grow into large open wounds. Some infections spread to vital organs and can become life threatening.

MRSA is a particular problem in hospitals where people with weakened immune systems are at greater risk. Young children and the elderly are also at particular risk. To reduce the spread of MRSA, surfaces and hands can be cleaned with antibacterials (commonly alcohol-based cleaning products).

MRSA and other antibiotic-resistant microorganisms **evolved** to be resistant to antibiotics because of the **misuse** or **overuse** of the antibiotic medicines themselves. That is, the more we prescribed and used them as medicines for humans and animals, the more opportunity the microorganisms had to evolve. This is an example of Darwin's theory of evolution by natural selection (see page 10).

Interdependence

Interdependence refers to a relationship between living organisms where organisms depend on each other for some resource or for survival. What happens to one organism will affect what happens to other organisms; everything is dependent on everything else. Interdependence is the dynamic relationship between all living things.

Food Chains

A **food chain** describes the feeding relationship between living organisms in an ecosystem. It also shows how **energy** and **biomass** are transferred along the food chain when the organisms feed. Biomass refers to the total mass (quantity) of organic material at each stage in the chain. Each stage in a food chain is called a **trophic level**. Lots of food chains in the same ecosystem can join up to make a **food web**.

Anything that eats another organism is called a **consumer**.

Grass (producer)   **Rabbit** (herbivore and primary consumer)   **Stoat** (carnivore and secondary consumer)   **Fox** (top carnivore and tertiary consumer)

Pyramids of Biomass

Energy enters a food chain from the **Sun**. Some energy and biomass are lost at each trophic level of a food chain as faeces (solid waste), movement energy and heat energy (especially by birds and mammals). This loss of energy at each trophic level limits the length of a food chain. Only a small amount of energy and biomass are incorporated into a consumer's body and transferred to the next feeding level. This is shown in a **pyramid of biomass**.

In the pyramid of biomass below, the loss of energy and biomass at each trophic level is indicated, and illustrates why such a representation of a food chain is always pyramid-shaped. This pyramid of biomass shows the food chain above quantitatively.

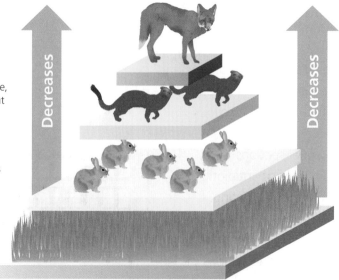

Energy

The fox gets the last tiny bit of energy left after all the others have had a share.

The stoats run around, mate, excrete, keep warm, etc. and pass on about $\frac{1}{10}$ of all the energy they got from the rabbits.

The rabbits run around, mate, excrete, keep warm, etc. and pass on about $\frac{1}{10}$ of all the energy they got from the grass.

The Sun is the energy source for all organisms, but only a fraction of the Sun's energy is actually captured in photosynthesis.

Decreases

Decreases

Biomass

The fox gets the biomass that remains to be passed on after all this.

The stoats lose quite a bit of biomass in droppings.

The rabbits lose quite a lot of biomass in droppings.

A lot of the biomass remains in the ground as the root system.

Parasitism and Mutualism

The survival of some organisms depends directly upon that of others. Examples of relationships like this can be split into two categories:

- **Parasitism** – where an infecting organism (parasite) benefits directly from its host (without killing it).
- **Mutualism** – where both organisms benefit from the relationship.

Examples of human parasites include **fleas**, **headlice** and **tapeworms**. **Mistletoe** is an example of a parasitic plant that grows on the branches of trees. It has green leaves for photosynthesis but takes water and minerals from its host plant.

Examples of mutualism are **oxpecker birds** that pick ticks and larvae from the hides of large mammals and **cleaner fish** that feed on the dead skin and parasites of larger fish.

> **HT** Other examples of mutualism include **nitrogen-fixing bacteria** that live in the roots of legumes (see page 28) and **chemosynthetic bacteria** that live in hydrothermal vents (see page 8). These bacteria form symbiotic relationships with tube worms.

Human Population and Pollution

The human population has experienced continuous growth for hundreds of years.

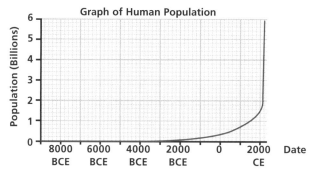

Graph of Human Population

Some scientists worry that the world is becoming overpopulated. They think that a number of the world's current problems, e.g. **extinction** of animals and plants, **global warming** and pollution of the air and water systems, are linked to this.

Pollution

Pollution is the contamination of the environment by **waste substances**, produced as the result of **human activity**.

Many waste substances are formed from the burning of fossil fuels to produce energy.

Air Pollution

Air pollution may consist of:

Hydrocarbons
• Released from combustion of fossil fuels (vehicles and factories).
• Cause smog, which can irritate skin and eyes.

Carbon Dioxide
• Released from combustion of fossil fuels (vehicles and factories).
• An important 'greenhouse gas'.

Sulfur Dioxide
• Released from combustion of fossil fuels (vehicles and factories).
• Contributes to acid rain.

Carbon Monoxide
• Released from vehicle exhausts and from many heavy industries.
• Poisonous gas.

Water Pollution

Water pollution may consist of:

Sewage (Human Waste)

- Bacteria feed on it and use up all the oxygen in the water. This causes other living things (e.g. water plants and fish) to die of asphyxiation (lack of oxygen).
- Contributes to eutrophication (see page 28).

Nitrates (Found in Fertilisers)

- Can be washed out of the soil into streams, rivers, ponds and lakes.
- Contributes to eutrophication.

Phosphates (Found in Waste Water from Laundries and Run-off from Fields)

- Constituent of fertiliser (as in NPK fertilisers – the P stands for phosphates).
- Contributes to eutrophication.

Investigating the Effect of Pollutants on Plant Growth

The effects of some water pollutants on plant growth can be seen by growing cress seeds in different solutions.

1. Place some seeds on damp cotton wool in three Petri dishes.
2. Apply a solution of ammonium nitrate and ammonium phosphate to the cotton wool in two different dishes.
3. Leave all three dishes in a sunny area, like a windowsill, for a few days.
4. Compare the growth of the seedlings in the two dishes with pollutants, with those in the third dish (which should have developed normally).

Indicator Species

Some animals and plants will only live in unpolluted areas. Their presence is an indicator for scientists of a clean environment and so they are called **indicator species** or bioindicators.

Examples of species that only live in clean water are the stonefly and freshwater shrimps.

Examples of air quality indicators are lichens and the blackspot fungus on roses. **Lichen** populations are very sensitive to sulfur dioxide air pollution and even quite low levels can kill them.

Few lichens indicate high concentration of sulfur dioxide in the air

Many lichens indicate low concentration of sulfur dioxide in the air

The presence of other species can indicate high levels of pollution. Blood worms and sludgeworms are present in polluted water.

Recycling

Recycling materials is now seen as a very important part of **sustainable development**, i.e. development that meets the needs of current generations without compromising the ability of future generations to meet their needs.

Local councils encourage us to recycle a range of materials including paper, plastic, metals and glass. They also encourage us to use compost heaps and bins for **biodegradable** items.

Recycling reduces demand for raw materials, reduces the problems of waste disposal and saves energy.

The Carbon Cycle

Carbon is an element that forms the basis of all living things. On Earth, the processes by which materials are removed by living things should, ideally, be balanced by processes that return them, so these materials can be recycled. The constant recycling of carbon is called the **carbon cycle**.

Plants and animals respire all the time. However, during daylight hours, plants also photosynthesise.

The amount of oxygen that plants use for respiration is only a tiny fraction of the amount they produce during photosynthesis. This means that the oxygen consumed by animals is more than adequately replaced by that produced by photosynthesis in normal circumstances. However, the levels of oxygen and carbon dioxide in the atmosphere depend upon the fine balance between respiration and photosynthesis being maintained.

Human activity, such as the burning of fossil fuels and deforestation, is starting to upset this balance.

The five main processes in the carbon cycle are as follows (see diagram below):

1. **Photosynthesis** – carbon dioxide is **removed** from the atmosphere by green plants to produce glucose. Some is returned to the air by the plants during respiration.
2. **Consumption** of plants by animals passes carbon compounds along food chains.
3. **Respiration** – plants and animals respire, **releasing** carbon dioxide into the atmosphere.
4. **Decay** – when plants and animals die, other animals and microorganisms feed on their bodies, causing them to break down, **releasing** carbon dioxide into the air. As the animals and microorganisms eat the dead plants and animals they respire, releasing carbon dioxide into the air.
5. **Combustion** – fossil fuels that are burned in power stations **release** carbon dioxide into the atmosphere.

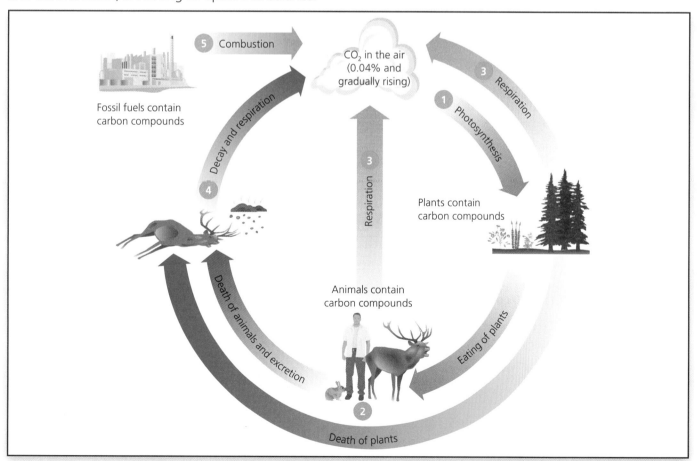

The Nitrogen Cycle

The **nitrogen cycle** shows how nitrogen and its compounds are recycled in nature. Nitrogen is a vital element for all living things and is used to make proteins, which are used in plant and animal growth. All enzymes are proteins. Nitrogen gas in the air cannot be used by plants and animals as it is inert (unreactive). Plants can only use it in the form of nitrates. The main processes in the nitrogen cycle are as follows:

1. **Nitrogen-fixing bacteria** convert atmospheric nitrogen into nitrates in soil. Some of these bacteria live in the soil, whilst others are found in the roots of leguminous plants in root nodules (e.g. pea plants).

2. When plants are eaten the nitrogen becomes animal protein and is passed along the food chain or web.

3. Dead organisms and waste contain ammonium compounds.

4. **Decomposers** break down dead animals and plants. **Soil bacteria** convert proteins and urea into ammonia.

5. **Nitrifying bacteria** convert ammonia into nitrates in the soil.

6. **Denitrifying bacteria** convert nitrates into atmospheric nitrogen.

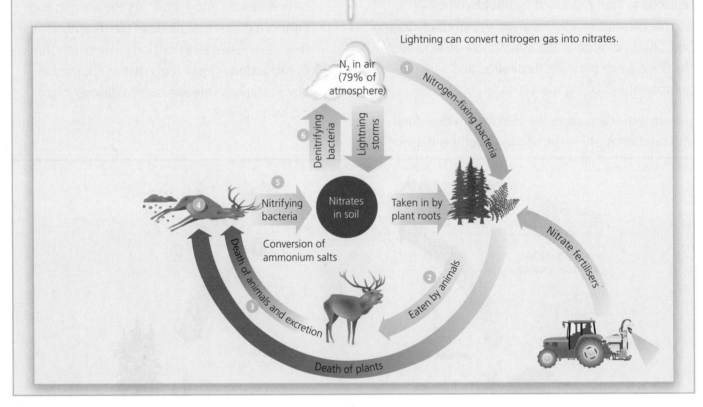

Lightning can convert nitrogen gas into nitrates.

N_2 in air (79% of atmosphere)

1 Nitrogen-fixing bacteria

6 Denitrifying bacteria

Lightning storms

5 Nitrifying bacteria

Nitrates in soil

Conversion of ammonium salts

Taken in by plant roots

Nitrate fertilisers

4

3 Death of animals and excretion

Death of plants

2 Eaten by animals

Eutrophication

Farmers use fertilisers to replace the nitrogen in the soil that has been used up by crops. This means that crop yields can be increased. But indiscriminate (careless) use of fertilisers can lead to environmental damage and **eutrophication**. Excess fertiliser is washed into streams and rivers and accumulates in ponds and lakes.

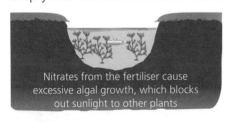

Nitrates from the fertiliser cause excessive algal growth, which blocks out sunlight to other plants

The other plants cannot photosynthesise so they die and start to rot

The rotting process uses up oxygen and the water cannot support life

Questions labelled with an asterisk (*) are ones where the quality of your written communication will be assessed – you should take particular care with your spelling, punctuation and grammar, as well as the clarity of expression, on these questions.

1 Living organisms are classified into many different groups.

*(a) Describe the characteristics of organisms in each of the five animal phyla. **(6)**

(b) State the difference between vertebrates and invertebrates. **(1)**

2 (a) Explain how fish and camels are adapted to their natural habitats. **(4)**

(b) State the definition of the term 'extreme habitats'. Give an example of an extreme habitat in your answer. **(2)**

(c) (i) State the term used to describe differences between individuals of the same species. **(1)**

(ii) State the two ways in which these differences can arise. **(2)**

(d) Variation within a species can be grouped into two types: continuous and discontinuous. State an example of each type in humans. **(2)**

3 *Describe Darwin's theory of natural selection. **(6)**

4 (a) Alternative forms of the same gene are called

A ☐ bases   C ☐ chromosomes

B ☐ DNA   D ☐ alleles   **(1)**

(b) (i) State a genetic disorder. **(1)**

(ii) Describe two of its symptoms. **(2)**

(c) Draw a genetic diagram to show the possible outcomes of a cross between a male with free earlobes (genotype EE) and a female with attached lobes (genotype ee). **(4)**

5 (a) State the definition of the term 'homeostasis'. **(1)**

(b) Explain the role of the skin in regulating our temperature when we become too hot. **(4)**

6 (a) The elongated parts of neurones are covered by an insulating layer called the

A ☐ nerve   C ☐ myelin sheath

B ☐ neurotransmitter D ☐ synapse   **(1)**

(b) The three types of neurone are sensory, relay and motor. Describe the function of each type. **(3)**

(c) Describe what happens at a synapse. **(4)**

7 (a) Describe how the body controls blood glucose levels when the concentration of blood glucose is too high. **(3)**

(b) Describe four symptoms of diabetes. **(4)**

(c) Describe the difference between Type 1 and Type 2 diabetes. **(2)**

(d) Suggest three things that diabetics could do on a daily basis, besides injecting insulin, to remain healthy. **(3)**

8 Plant shoots growing towards the light is called

A ☐ positively geographic   **C** ☐ positively gravitropic

B ☐ positively phototropic   **D** ☐ negatively phototropic   **(1)**

9 **(a)** State the definition of the term 'drugs'. **(1)**

(b) Describe the effects that stimulants and depressants have upon the nervous system.
Give an example of each in your answer. **(4)**

10 **(a)** State three organs or tissues that can be transplanted. For each one, state whether it is an organ or a tissue. **(3)**

(b) Suggest three of the ethical issues associated with organ transplants. **(3)**

11 **(a)** State the three types of pathogen. **(3)**

(b) State the definition of the term 'vectors of disease'. Give an example in your answer. **(2)**

(c) State two physical and two chemical barriers that stop animals being attacked by pathogens. **(4)**

12 **(a)** State the definition of the term 'interdependence'. **(1)**

(b) Draw a food chain with four trophic levels. Label each stage with appropriate key words to identify the types of organisms present. **(4)**

(c) Describe how the levels in a pyramid of biomass are different. **(1)**

(d) Describe the difference between parasitic and mutualistic relationships.
Give an example of each type in your answer. **(4)**

13 State two substances that can pollute air and two substances that can pollute water.
For each substance state where it comes from. **(4)**

14 *Describe the main processes in the carbon cycle. **(6)**

HT

15 Explain why classification is important. **(2)**

16 Describe how the genetic disorder cystic fibrosis is inherited. **(1)**

17 **(a)** Explain how vasoconstriction helps mammals to retain heat. **(1)**

(b) Explain how vasodilation helps mammals to lose heat. **(1)**

18 Describe how the body controls blood glucose levels when the concentration of blood glucose is too low. **(3)**

19 Describe three commercial uses of plant hormones. **(3)**

20 **(a)** Describe the medical significance of MRSA. **(1)**

(b) Explain why MRSA have evolved. **(1)**

21 *Describe the main processes in the nitrogen cycle. Highlight the importance of bacteria. **(6)**

C1 Topic 1: The Earth's Sea and Atmosphere

This topic looks at:

- how the early atmosphere and oceans were formed
- how levels of carbon dioxide reduced and levels of oxygen increased
- the impact of volcanic and human activities on the atmosphere

History of the Atmosphere

Since the formation of the Earth 4.6 billion years ago, the atmosphere has changed dramatically.

The timescale, however, is enormous: one billion years is one thousand million (1 000 000 000) years.

Timescale	Composition of the Atmosphere	Key Factors and Events that Shaped the Atmosphere
Formation of the Earth 4 billion years ago 3 billion years ago 2 billion years ago 1 billion years ago now	Water Vapour and Other Gases / Carbon dioxide / Carbon dioxide much reduced / Increase in oxygen and nitrogen / 1% Other Gases / CO₂ / Nitrogen 78% / Oxygen 21%	Volcanic activity released carbon dioxide and small amounts of ammonia, methane and hydrogen. This volcanic activity also produced water vapour, which condensed as the Earth cooled, falling as rain and filling up the hollows in the crust to form oceans. Primitive green plants such as single-celled algae evolved and the amount of carbon dioxide was reduced as the plants took it in and gave out oxygen during photosynthesis. Carbon from carbon dioxide in the air was locked up in sedimentary rocks as carbonates and fossil fuels. The carbonates were formed when carbon dioxide dissolved in the oceans (see page 32). There is now more oxygen in the atmosphere. Some is converted to ozone, protecting living organisms from the Sun's ultraviolet radiation and allowing new species to evolve. There is a state of approximate balance because: • photosynthesis produces oxygen in the presence of sunlight and uses carbon dioxide • respiration and burning fuels use oxygen and produce carbon dioxide • carbon dioxide is absorbed by the seas and oceans.

Scientific Evidence About the History of the Atmosphere

Scientists have different sources of information about the development of the Earth's atmosphere, which means they are unable to be precise about the evolution of the Earth's atmosphere.

Scientists do have the following evidence, which has helped them to conclude that there was little or no oxygen in the early atmosphere and that the amount of oxygen within the atmosphere has increased over time:

* Oxygen is not released by volcanoes.
* Venus and Mars have similar atmospheres made up of carbon dioxide.
* Fossil evidence from early rock shows the existence of photosynthesising organisms.
* The composition of iron compounds is different within different types and ages of rock.

Carbon Dioxide in the Atmosphere

When the Earth was young, the atmosphere was mostly made up of carbon dioxide. Gradually this decreased due to photosynthesis of primitive plants converting the carbon dioxide and water to glucose and oxygen. The amount of carbon dioxide in the atmosphere also decreased as it dissolved in the newly formed oceans.

Some of the dissolved carbon dioxide reacted with other substances in the sea to form insoluble compounds such as **calcium carbonate** and soluble compounds such as **calcium hydrogencarbonate**. These compounds became concentrated in the shells of sea creatures. When the sea creatures died their remains eventually formed carbonate rocks, locking away the carbon.

The levels of carbon dioxide in the atmosphere are generally thought to have been in balance over the last 200 million years.

The constant recycling of carbon is called the **carbon cycle.**

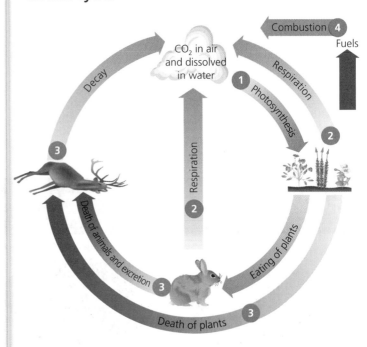

1. Carbon dioxide is removed from the atmosphere by green plants during photosynthesis.
2. Plants and animals respire, releasing carbon dioxide into the atmosphere.
3. Microorganisms feed on dead plants and animals, causing them to break down, decay and release carbon dioxide into the air. (The microorganisms respire as they feed.)
4. The burning of fossil fuels also releases carbon dioxide into the air.

Present day volcanic eruptions contribute to the levels of carbon dioxide within the atmosphere. The burning of fossil fuels also releases carbon dioxide into the atmosphere. If any branch of the carbon cycle is changed, then this will impact on the level of carbon dioxide in the atmosphere, resulting in either an increase or a reduction.

For example, deforestation takes away the trees that convert the carbon dioxide into oxygen, so there will be an increase in levels of carbon dioxide.

Animals exhale carbon dioxide when they respire, so an increase in farming livestock increases the amount of carbon dioxide produced through respiration.

Global Warming

Global warming refers to the gradual increase in the Earth's temperature that has occurred. This can happen when greenhouse gases, such as carbon dioxide, water vapour, nitrous oxide and methane, trap energy from the Sun in the Earth's atmosphere, which increases the temperature. The level of greenhouse gases in the atmosphere has gradually increased as a result of:

- **burning fossil fuels** (oil, coal, gas), which produce carbon dioxide, sulfur dioxide (which contributes to **acid rain**) and carbon monoxide
- motorised **transportation** burning petrol and diesel, which produces carbon dioxide, sulfur dioxide and carbon monoxide
- an increase in **cattle farming and rice growing** – methane is released from wetlands (where rice grows) and from animals (particularly cattle)
- **deforestation** – when trees grow, they take in carbon dioxide. If more and more trees are cut down, less carbon dioxide is removed from the atmosphere.

Oxygen in the Atmosphere

The levels of oxygen in the atmosphere have changed over the history of the Earth, rising steadily from non-existence about 2 billion years ago, to about 21% today. Scientists believe it was the evolution of primitive plants such as algae that caused the atmosphere to change and the carbon dioxide to be used up during photosynthesis. This process produced the first molecules of oxygen.

Carbon dioxide	+	Water	**Light** →	Glucose	+	Oxygen

(HT) $6CO_2$ + $6H_2O$ $\xrightarrow{\text{Light}}$ $C_6H_{12}O_6$ + $6O_2$

Measuring Oxygen in the Atmosphere

A simple practical demonstration can be carried out to show the level of oxygen present in the atmosphere.

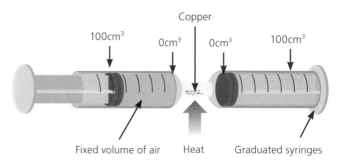

Fixed volume of air   Heat   Graduated syringes

If copper is heated while a fixed volume of air passes over it, then the volume of air will gradually reduce as the oxygen is used up in the chemical reaction. Eventually the volume of air will stop changing because all the oxygen will have been used up.

Copper	+	Oxygen	→	Copper oxide

(HT) $2Cu_{(s)}$ + $O_{2(g)}$ → $2CuO_{(s)}$

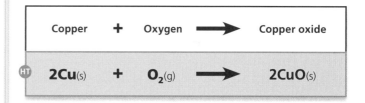

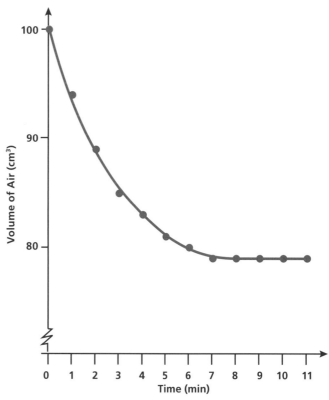

Volume of Air during the Course of the Experiment

Different Types of Rock

Rocks are mixtures of **minerals**. A mineral is any solid element or compound that can be found in the Earth's crust.

There are three types of rock:
- igneous
- sedimentary
- metamorphic.

Igneous rock is produced when molten rock cools down to form crystals.

Molten rock, or magma, that erupts from a volcano is called lava. When lava cools, it cools very quickly, forming small crystals. This type of rock is called **extrusive igneous rock**. An example of this type of rock is basalt.

If the magma does not reach the surface it can still cool down but will do so very slowly, forming large crystals. This type of rock is called **intrusive igneous rock**. An example of this type of rock is granite.

Granite

Sedimentary rock, such as chalk and limestone, is formed as a result of a build-up of layers of sediment.

The pressure (compaction) of the layers forces out water but leaves behind the minerals that had been dissolved in it. These minerals act like a cement and hold the particles together. The layers that make up sedimentary rock are where **fossils** can be found.

However, **sedimentary rock** is porous because there are tiny gaps between the particles. These tiny gaps make the rock susceptible to erosion because they are points of weakness.

Metamorphic rock forms as a result of rock being buried due to the movement of the Earth's crust. The deeper the rock is buried, the greater the heat and pressure to which it is exposed. Over time this will cause the rock to change without ever melting. The minerals in the rock will line up to form bands or sheets of tiny grains or crystals. If this happens to limestone or chalk then eventually marble is formed.

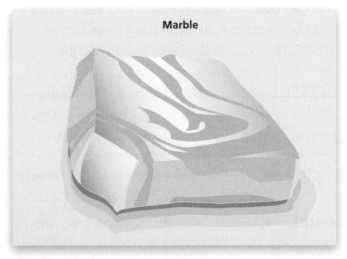

Marble

Limestone

Calcium carbonate, $CaCO_3$, can be found in rocks that exist in the Earth's crust as chalk, limestone and marble. Limestone is a very important building material and has industrial uses, for example, it is used in the extraction of iron from its ore (see page 41).

Quarrying Limestone

To extract limestone from the ground it must be **quarried**. This means digging a large hole and taking the rock away.

There are many factors that have to be considered before the limestone can be quarried. These factors include:

- the effect on the landscape and native animal habitats
- the effect on local businesses
- costs involved in quarrying the rock and then processing it
- whether there is an available local workforce
- the effect of noise pollution
- if there will be an increase in air dust and dirt levels
- if there will be additional traffic and new roads required
- how the quarry site can be used afterwards.

Uses of Limestone

As a building material, limestone is used in numerous ways.

It can be used straight from the quarry as block to construct the walls and floors of buildings. Many buildings built near limestone quarries are manufactured in this way.

Glass is made by heating limestone, $CaCO_3$, with sand, SiO_2, and sodium carbonate, Na_2CO_3. When they are heated they all melt together and then cool to form a transparent solid.

Cement is the result of heating powdered limestone and clay together in a rotary kiln. It is the heat that causes the two materials to react together.

Concrete is the result of mixing cement, sand, crushed rock or gravel and water together. As the mixture sets it forms a hard, stone-like solid. Steel supports can be used to make the concrete stronger. This is called reinforced concrete.

Acidic sulfur dioxide gas emitted by coal-fired power stations can be neutralised by using limestone or lime before the gas leaves the chimneys.

This reaction forms calcium sulfate, $CaSO_4$, which can be used in the building industry as **plaster**.

Farmers can **neutralise soil** that is too acidic to grow crops by spreading powdered limestone or lime onto the soil. Lime is another name for calcium oxide, CaO. Calcium oxide can be obtained from the thermal decomposition of calcium carbonate.

Thermal Decomposition of Carbonates

When calcium carbonate is heated strongly enough, it will break down and form different molecules. This process is called **thermal decomposition**. Thermal decomposition is very important for calcium carbonate as it produces calcium oxide, a very reactive material that has many uses, including in the neutralisation of acidic soils, as indicated on the previous page.

The equation below describes the thermal decomposition of calcium carbonate.

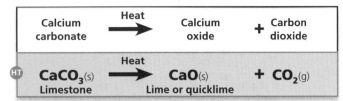

$$CaCO_3{(s)} \xrightarrow{Heat} CaO{(s)} + CO_2{(g)}$$
Limestone — Lime or quicklime

The following equations show how limewater (used when testing for carbon dioxide gas) is produced from calcium oxide.

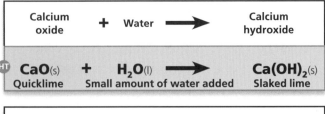

$$CaO{(s)} + H_2O{(l)} \longrightarrow Ca(OH)_2{(s)}$$
Quicklime — Small amount of water added — Slaked lime

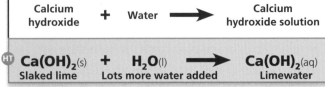

$$Ca(OH)_2{(s)} + H_2O{(l)} \longrightarrow Ca(OH)_2{(aq)}$$
Slaked lime — Lots more water added — Limewater

Some carbonates will decompose more easily than others and this can be demonstrated in a simple investigation in the laboratory.

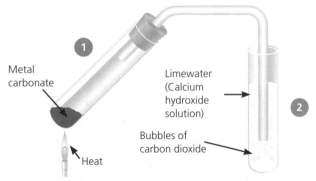

The more bubbles that appear in the limewater, the easier the carbonate is to decompose.

Bubbling the carbon dioxide through limewater means it will not all escape. It will also result in the limewater turning milky white or cloudy as calcium carbonate is precipitated.

For example, the equations below illustrate what happens during the decomposition of zinc carbonate:

Stage ①

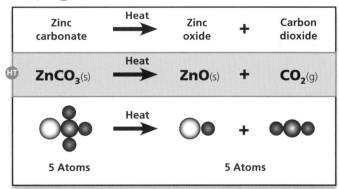

$$ZnCO_3{(s)} \xrightarrow{Heat} ZnO{(s)} + CO_2{(g)}$$

5 Atoms — 5 Atoms

Stage ②

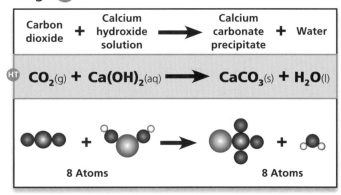

$$CO_2{(g)} + Ca(OH)_2{(aq)} \longrightarrow CaCO_3{(s)} + H_2O{(l)}$$

8 Atoms — 8 Atoms

From the equation it can be seen that no atoms have been lost or gained, they have just rearranged themselves. **Conservation of mass** means that total masses at the start and finish will also have stayed the same.

Remember that **atoms** are the basic particles from which all matter is made. All chemical **elements** are made up of atoms.

Balancing Equations

All chemical reactions follow the same simple rule: the total mass of the reactants is equal to the total mass of the products.

This means there must be the same number of atoms on both sides of the equation.

Writing Balanced Equations

1. Write down the word equation for the reaction.
2. Write down the correct formula for each of the reactants and the products.
3. Check that there are the same numbers of each atom on both sides of the equation.

If the equation is already balanced, leave it.

If the equation needs balancing:

4. Write a number in front of one or more of the formulae. This increases the number of all of the atoms in the formula.
5. Don't forget the state symbols: (s) = solid, (l) = liquid, (g) = gas and (aq) = dissolved in water (aqueous solution).

Example

Balance the reaction between sodium and water.

1. The word equation for the reaction is:

$$\text{Sodium} + \text{Water} \longrightarrow \text{Sodium hydroxide} + \text{Hydrogen}$$

2. The correct formulae for each of the products and the reactants are:

$$Na + H_2O \longrightarrow NaOH + H_2$$

3. Check there are the same number of atoms on each side of the equation.

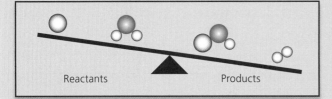

Reactants   Products

There are more hydrogen atoms on the products side than on the reactants side. The equation needs balancing.

4. Balance the hydrogen by doubling the amount of water and sodium hydroxide.

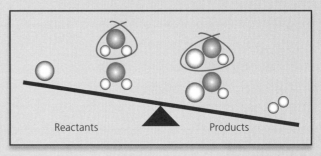

Reactants   Products

The amount of oxygen and hydrogen on both sides is equal. However, the amount of sodium is now unequal.

Double the sodium on the reactants side to match the amount of sodium on the products side.

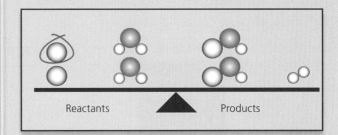

Reactants   Products

This gives you the balanced equation.

$$2Na + 2H_2O \longrightarrow 2NaOH + H_2$$

5. Don't forget to include the state symbols.

$$2Na_{(s)} + 2H_2O_{(l)} \longrightarrow 2NaOH_{(aq)} + H_{2(g)}$$

C1 Topic 3: Acids

This topic looks at:

- the importance of neutralisation reactions and the electrolysis of hydrochloric acid
- the neutralisation reactions of other acids
- the uses of chlorine and methods of producing it
- what happens during the electrolysis of water

Acids

An **acid** is a substance that produces **hydrogen ions** when in an aqueous solution, for example, hydrochloric acid (HCl), sulfuric acid (H_2SO_4) and nitric acid (HNO_3). Hydrogen ions can be identified by using an **indicator**, such as a universal indicator, which indicates a range of pH values.

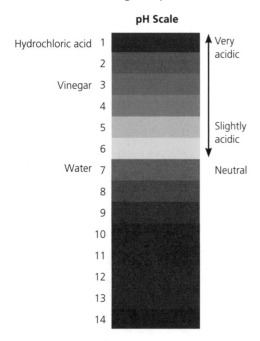

pH Scale

Hydrochloric acid	1	Very acidic
	2	
Vinegar	3	
	4	
	5	Slightly acidic
	6	
Water	7	Neutral
	8	
	9	
	10	
	11	
	12	
	13	
	14	

Making Salts

A neutral **salt** can be formed when an acid is reacted with a base. (Some bases are soluble in water and dissolve to produce alkaline solutions.) This type of reaction is called a **neutralisation** reaction. Most of the salts formed in these reactions are soluble and they can only be obtained when the water is evaporated off.

Using an Alkaline Hydroxide Base

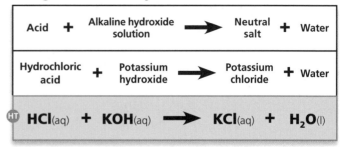

Acid	+	Alkaline hydroxide solution		Neutral salt	+	Water
Hydrochloric acid	+	Potassium hydroxide		Potassium chloride	+	Water

HT $HCl_{(aq)} + KOH_{(aq)} \longrightarrow KCl_{(aq)} + H_2O_{(l)}$

Both beakers must contain the same number of acidic hydrogen (H^+) ions and alkaline hydroxide (OH^-) ions if they are to neutralise each other exactly.

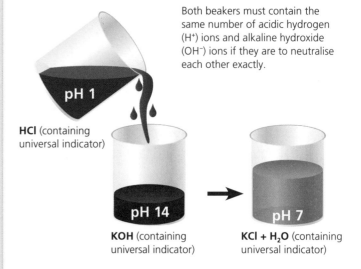

HCl (containing universal indicator)

pH 1

KOH (containing universal indicator)

pH 14

$KCl + H_2O$ (containing universal indicator)

pH 7

Using a Metal Oxide Base

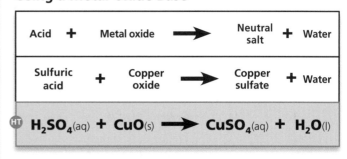

Acid	+	Metal oxide		Neutral salt	+	Water
Sulfuric acid	+	Copper oxide		Copper sulfate	+	Water

HT $H_2SO_4{}_{(aq)} + CuO_{(s)} \longrightarrow CuSO_4{}_{(aq)} + H_2O_{(l)}$

Using a Metal Carbonate Base

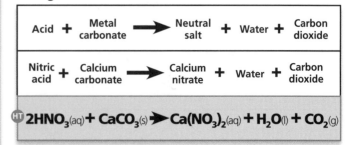

Acid	+	Metal carbonate		Neutral salt	+	Water	+	Carbon dioxide
Nitric acid	+	Calcium carbonate		Calcium nitrate	+	Water	+	Carbon dioxide

HT $2HNO_3{}_{(aq)} + CaCO_3{}_{(s)} \longrightarrow Ca(NO_3)_2{}_{(aq)} + H_2O_{(l)} + CO_2{}_{(g)}$

The particular salt produced depends on the acid and the metal in the base used. They all react in the same way but produce different salts. Sulfuric acid produces sulfate salts, hydrochloric acid produces chloride salts and nitric acid produces nitrate salts.

Everyday Neutralisation Reactions

Hydrochloric acid is used by the human body to digest food. It is an acid that is produced in the stomach to help break down food into smaller molecules. It will also kill any germs or bacteria that may have been present on the ingested food.

Sometimes too much acid is produced by the stomach and this can cause pain. Excess acid in the stomach can be neutralised by taking indigestion tablets. Each of these tablets or remedies contains a base that will neutralise the acid.

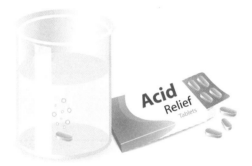

Remember:

| Acid | + | Alkali | ⟶ | Salt | + | Water |

This can be tested in the laboratory by neutralising some hydrochloric acid with a range of different indigestion remedies.

Electrolysis

When electrical energy from a direct current (d.c.) supply is passed through a compound that is molten or in solution, the compound is **decomposed**. The ions move to the electrode of opposite charge.

The ions that are positively charged move towards the negative electrode.

The ions that are negatively charged move towards the positive electrode.

When the ions get to the electrodes, they are **discharged** (lose their charge) and atoms of elements are formed.

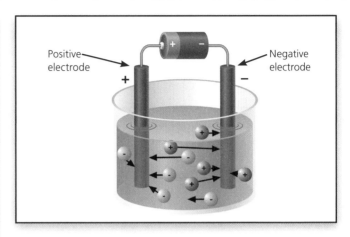

Positive electrode
Negative electrode

Electrolysis of Hydrochloric Acid

Electrolysis can be demonstrated in the laboratory by passing a current through dilute hydrochloric acid.

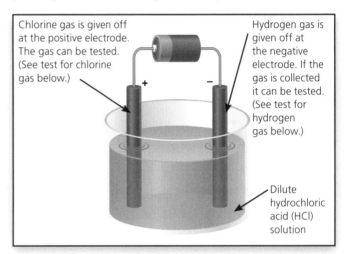

Chlorine gas is given off at the positive electrode. The gas can be tested. (See test for chlorine gas below.)

Hydrogen gas is given off at the negative electrode. If the gas is collected it can be tested. (See test for hydrogen gas below.)

Dilute hydrochloric acid (HCl) solution

The following table shows how to test for hydrogen gas and chlorine gas.

Gas	Properties	Test for Gas
Hydrogen, H_2	A colourless gas. It combines violently with oxygen when ignited.	When mixed with air, burns with a squeaky pop.
Chlorine, Cl_2	A green poisonous gas that bleaches dyes.	Turns damp indicator paper white.

Electrolysis of Water

The electrolysis of water will result in its decomposition to hydrogen gas at the negative electrode and oxygen gas at the positive electrode.

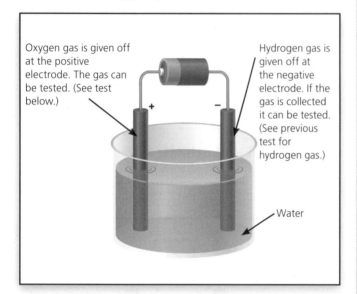

Oxygen gas is given off at the positive electrode. The gas can be tested. (See test below.)

Hydrogen gas is given off at the negative electrode. If the gas is collected it can be tested. (See previous test for hydrogen gas.)

Water

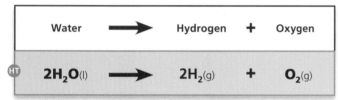

| Water | → | Hydrogen | + | Oxygen |

HT $2H_2O_{(l)} \longrightarrow 2H_{2(g)} + O_{2(g)}$

The presence of these gases can be tested by using the test described on page 39 for hydrogen gas and the test for oxygen shown below.

Gas	Properties	Test for Gas
Oxygen, O_2	A colourless gas that helps fuels burn more readily than in air.	Relights a glowing splint.

The Production of Chlorine by Electrolysis

Chlorine gas can be obtained during the industrial electrolysis of brine. Brine is the name that is given to water containing large amounts of salt (sodium chloride), such as sea water. Extra safety precautions must be taken by a company when chlorine gas is obtained on such a large industrial scale because it is a toxic gas.

When a concentrated solution of sodium chloride is electrolysed, chlorine gas is produced at the positive electrode, hydrogen gas is produced at the negative electrode and the solution that remains is the alkali sodium hydroxide.

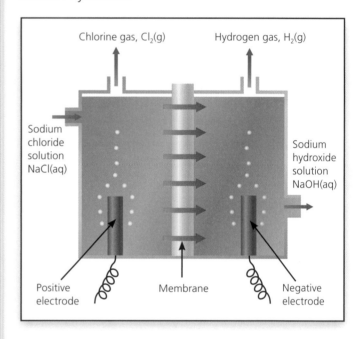

Chlorine gas, $Cl_2(g)$

Hydrogen gas, $H_2(g)$

Sodium chloride solution NaCl(aq)

Sodium hydroxide solution NaOH(aq)

Positive electrode

Membrane

Negative electrode

| Sodium chloride | + Water → | Hydrogen | + Chlorine | + | Sodium hydroxide |

HT $2NaCl_{(aq)} + 2H_2O_{(l)} \longrightarrow H_{2(g)} + Cl_{2(g)} + 2NaOH_{(aq)}$

Uses of Chlorine

Chlorine is a toxic gas that will kill harmful bacteria when added to water. It can be used to make bleaches and other disinfectants and can also be used to make the polymer poly(chloroethene) (PVC) (see page 50).

C1 Topic 4: Obtaining and Using Metals

This topic looks at:
- how metals are obtained from their ores
- what reduction and oxidation reactions are
- how metals are disposed of
- how properties determine the use of a metal
- how the reactivity series can predict extraction methods and reactions of metals
- how the properties of an alloy differ from the original metal

Metal Ores

Ores are naturally occurring rocks found in the Earth's crust. They contain **compounds** of metals in sufficient amounts to make it worthwhile extracting them.

Some of these compounds are **metal oxides**, for example, iron oxide (haematite). The method of extracting a metal from its ore depends on the metal's position in the **reactivity series**.

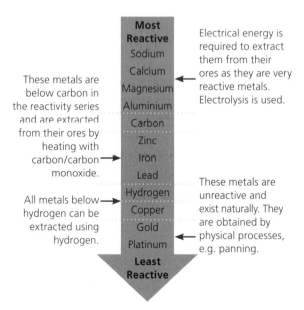

Reactivity Series

These metals are below carbon in the reactivity series and are extracted from their ores by heating with carbon/carbon monoxide.

All metals below hydrogen can be extracted using hydrogen.

Most Reactive
Sodium
Calcium
Magnesium
Aluminium
Carbon
Zinc
Iron
Lead
Hydrogen
Copper
Gold
Platinum
Least Reactive

Electrical energy is required to extract them from their ores as they are very reactive metals. Electrolysis is used.

These metals are unreactive and exist naturally. They are obtained by physical processes, e.g. panning.

The most reactive metals form the most stable ores and are therefore most difficult to extract.

The least reactive metals are found uncombined in the Earth's crust and are the easiest to extract from their ores.

Electrolysis of Molten Aluminium Oxide

Aluminium is extracted from bauxite (impure aluminium oxide). It takes a lot of energy to melt bauxite so it is first mixed with another ore of aluminium called **cryolite** as this will reduce the melting point of the bauxite. Bauxite has to be molten before it can be electrolysed to produce aluminium.

During this process:

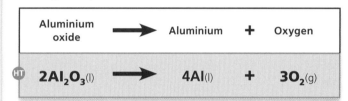

| Aluminium oxide | → | Aluminium | + | Oxygen |

HT $$2Al_2O_{3(l)} \longrightarrow 4Al_{(l)} + 3O_{2(g)}$$

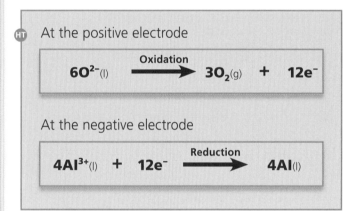

HT At the positive electrode

$$6O^{2-}_{(l)} \xrightarrow{\text{Oxidation}} 3O_{2(g)} + 12e^-$$

At the negative electrode

$$4Al^{3+}_{(l)} + 12e^- \xrightarrow{\text{Reduction}} 4Al_{(l)}$$

Extraction of Iron

To extract iron from its ore, **haematite** (iron(III) oxide) is smelted with carbon (in the form of coke) and limestone. (Limestone is only used to get rid of the waste.) This is a combination of reduction and oxidation processes (see page 42) that takes place inside the **blast furnace**.

The iron oxide is reduced to iron by carbon monoxide gas that is produced by the smelting processes.

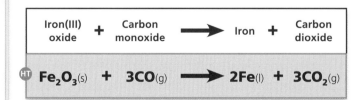

| Iron(III) oxide | + | Carbon monoxide | → | Iron | + | Carbon dioxide |

HT $$Fe_2O_{3(s)} + 3CO_{(g)} \longrightarrow 2Fe_{(l)} + 3CO_{2(g)}$$

Reduction

Reduction is the **loss of oxygen** from a compound during a chemical reaction. It is the process through which a metal compound is broken down to give the metal element. For example:

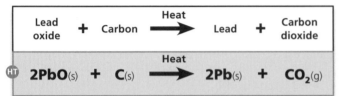

HT
$$2PbO_{(s)} + C_{(s)} \xrightarrow{Heat} 2Pb_{(s)} + CO_{2(g)}$$

Demonstrating Reduction of a Metal Oxide

By mixing together a spatula of carbon and copper oxide and then heating them strongly, a pinkish brown powder of copper will be formed.

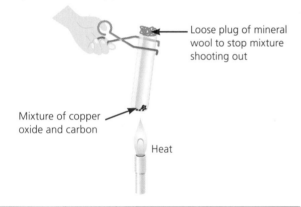

Loose plug of mineral wool to stop mixture shooting out

Mixture of copper oxide and carbon

Heat

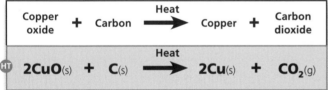

HT
$$2CuO_{(s)} + C_{(s)} \xrightarrow{Heat} 2Cu_{(s)} + CO_{2(g)}$$

Oxidation

When a metal or any other element combines with oxygen to make another substance, it is called **oxidation**. Oxidation is the **gaining of oxygen** by an element. For example, when magnesium is heated with oxygen it is oxidised to produce magnesium oxide.

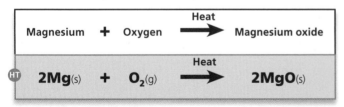

HT
$$2Mg_{(s)} + O_{2(g)} \xrightarrow{Heat} 2MgO_{(s)}$$

Corrosion of Metals

Metals are able to corrode. **Corrosion** happens when the metal reacts with oxygen in the air. The metal becomes oxidised because it has combined with oxygen to form the metal oxide.

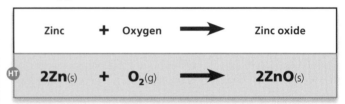

HT
$$2Zn_{(s)} + O_{2(g)} \longrightarrow 2ZnO_{(s)}$$

Reactivity Series

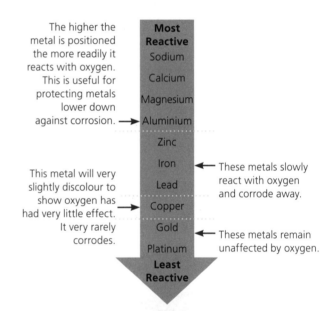

The higher the metal is positioned the more readily it reacts with oxygen. This is useful for protecting metals lower down against corrosion. → Aluminium

Most Reactive
Sodium
Calcium
Magnesium
Aluminium
Zinc
Iron
Lead
Copper
Gold
Platinum
Least Reactive

These metals slowly react with oxygen and corrode away.

This metal will very slightly discolour to show oxygen has had very little effect. It very rarely corrodes.

These metals remain unaffected by oxygen.

Corroded Car Made of Iron

Uncorroded Car Made of Aluminium

Recycling Metals

When we **recycle** something that we no longer need, we make a new product from the old material and at the same time we are able to conserve our planet's resources.

More and more materials can be recycled but the most common ones are metals, paper and glass.

When metals are recycled less energy is needed to process them into a new product as there are no impurities to get rid of. Aluminium and steel are currently the easiest metals to recycle.

Economic and Environmental Considerations

Recycling makes environmental sense because it:
- saves on raw materials
- saves on landfill sites and costs associated with waste disposal
- needs less energy to process the material than when the substance was first produced, so it saves on fossil fuels
- cuts down on excavation and mining so there is less environmental damage and fewer waste products
- uses less water and chemicals so reduces pollution.

Industries will tend to invest in recycling if it saves them money.

Some resources can be sustained over a long period by managing the resource, e.g. planting a tree for every tree that is cut down.

Sustainable development involves helping people to satisfy their basic requirements and enjoy a good standard of living without compromising future generations.

Properties and Uses of Metals

Aluminium, copper and gold are examples of **pure** metals. A pure metal contains atoms of that element only. Most metals extracted from the ground need to be purified.

When metals have been purified they will show many of the following general properties:
- good conductivity (they make good conductors of heat and electricity)
- dense (feel heavy)
- malleable (can be hammered into shape without cracking)
- shiny when polished
- high melting points
- strong under tension and compression
- sonorous (will ring when hit)
- ductile (can be drawn into a wire).

The following table shows the properties and uses of three pure metals.

Metal	Uses	Property
Aluminium	High-voltage power cables, furniture, drinks cans, foil food wrap	Corrosion resistant, ductile, malleable, good conductivity, low density
Copper	Electrical wiring, water pipes, saucepans	Ductile, malleable, good conductivity
Gold	Jewellery, electrical junctions	Ductile, shiny, good conductivity

Metals can also be alloyed. An **alloy** is a mixture of metals, usually produced to make the original metal stronger or improve its resistance to corrosion.

Different metals have atoms of different sizes. In an alloy, the mixture of sizes makes it harder for layers of metal ions to slide over each other, so the alloy is stronger than the individual pure metals. In a similar way, **conglomerate** rock is stronger than sandstone. Conglomerate contains grains and pebbles of various sizes, whereas sandstone contains grains of roughly the same size.

Alloys

Many metals are far more useful when they are **not pure**. Mixing a metal with other metals can change the properties of the metal.

The resulting alloy has a greater range of uses than the original metal. Remember, an alloy is a **mixture**. An alloy can have:

- a lower melting point (useful for solder)
- increased corrosion resistance (useful for anything that will be exposed to air and water)
- increased chemical resistance (useful for storing chemicals)
- increased strength and hardness (useful in construction of bridges, aircraft, cars, etc.).

Examples of Using an Alloy

Pure iron is not good for building things because it is too soft and bends easily. It also corrodes easily. If a small amount of carbon is present in iron, mild steel is produced, which is hard and strong, so it can be used for building things.

Steel is an example of an alloy of iron.

Mild Steel	99.5% Fe, 0.5% C	Hard but easily worked
Hard Steel	99% Fe, 1% C	Very hard so higher strength but brittle
Duriron	84% Fe, 1% C, 15% Si	Not affected by acid

If nickel and chromium are mixed with iron, stainless steel is produced, which is hard and rustproof, so can be used in areas that are exposed to air and water, e.g. for cutlery.

⊞ Nitinol

Nitinol is a mixture of nickel and titanium. It is also an example of a smart or shape memory alloy because it will return to its original form after any stress has been released. For example, nitinol will return to its original shape after it has been heated in hot water or an electrical current has been passed through it. It is mostly used in medical applications, such as spectacle frames and stents (supports) in broken or damaged blood vessels.

Pure Gold

Pure gold is a very soft metal. Gold is too soft for many applications because it can be damaged easily. To make it stronger, gold is alloyed with other metals. For yellow gold applications, it is alloyed with copper, silver and zinc. For white gold applications, it is alloyed with nickel, copper and zinc or palladium, copper and silver.

The purity of gold is measured in carats and fineness. A carat is one way of showing the proportion of gold present in an alloy, on a scale of 1 to 24. For example, 100% pure gold would be classified as 24 carat, an alloy containing 75% pure gold would be classified as 18 carat and an alloy containing 50% pure gold would be classified as 12 carats.

Fineness is measured with a three-figure numbering system. This is also seen in hallmarking of gold items. Purest gold is marked as 999 and 75% gold is marked as 750.

C1 Topic 5: Fuels

This topic looks at:
- what crude oil is and what it can be separated into
- how properties of fractions differ
- complete and incomplete combustion and the associated dangers
- the advantages and disadvantages of biofuels and hydrogen cells
- how some polymers are made, and their everyday uses and problems
- equations of polymerisation reactions

Crude Oil and Hydrocarbons

Crude oil is a mixture of **hydrocarbons** (i.e. compounds containing only hydrogen and carbon). The properties of the different hydrocarbons in crude oil remain unchanged and specific. This makes it possible to separate them by **fractional distillation**.

The oil is evaporated by heating, then allowed to condense at a range of different temperatures to form fractions. Each resulting fraction contains hydrocarbon molecules with a similar number of carbon atoms. The process takes place in a **fractionating column**.

As hydrocarbons get larger (i.e. the greater the number of carbon atoms in a molecule) they become:
- more viscous
- less flammable
- less volatile

and they have a higher boiling point.

Hydrocarbon	No. of Carbon Atoms
Refinery gases	1–4
Gasoline	5–6
Kerosene	10–16
Diesel oil	15–22
Fuel oil	30–40
Bitumen	50+

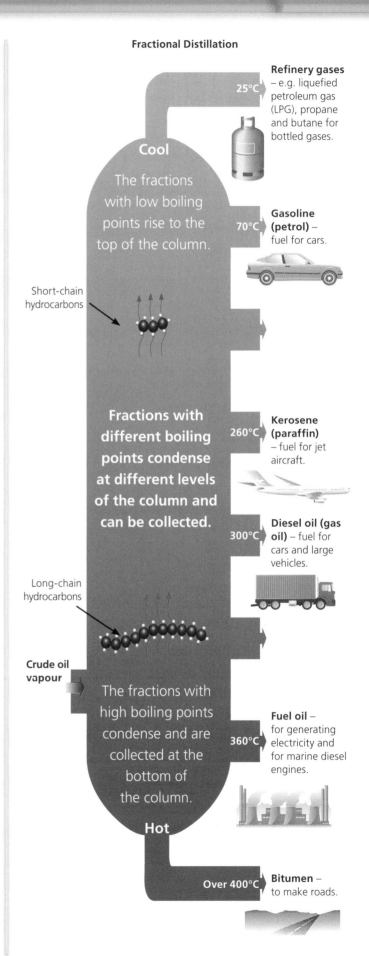

Fractional Distillation

Cool

The fractions with low boiling points rise to the top of the column.

Short-chain hydrocarbons

Fractions with different boiling points condense at different levels of the column and can be collected.

Long-chain hydrocarbons

Crude oil vapour

The fractions with high boiling points condense and are collected at the bottom of the column.

Hot

25°C — **Refinery gases** – e.g. liquefied petroleum gas (LPG), propane and butane for bottled gases.

70°C — **Gasoline (petrol)** – fuel for cars.

260°C — **Kerosene (paraffin)** – fuel for jet aircraft.

300°C — **Diesel oil (gas oil)** – fuel for cars and large vehicles.

360°C — **Fuel oil** – for generating electricity and for marine diesel engines.

Over 400°C — **Bitumen** – to make roads.

Combustion

A **fuel** is a substance that releases useful amounts of energy when burned. Many fuels are hydrocarbons, e.g. methane from natural gas.

When a fuel burns it reacts with oxygen from the air.

Fuels burn with different flame colours, e.g. coal burns with a dirty yellow flame and methylated spirits burn with a purple flame.

Complete Combustion

When a hydrocarbon burns and there is plenty of oxygen available, complete combustion occurs, producing carbon dioxide (which can be can tested for by bubbling through limewater) and water, and releasing **energy**.

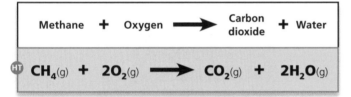

Methane + Oxygen ⟶ Carbon dioxide + Water

HT $CH_{4(g)} + 2O_{2(g)} \longrightarrow CO_{2(g)} + 2H_2O_{(g)}$

Incomplete Combustion

Sometimes a fuel burns without sufficient oxygen, e.g. in a room with poor ventilation. Then, incomplete combustion takes place. Instead of carbon dioxide being produced, carbon monoxide is formed.

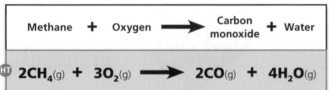

Methane + Oxygen ⟶ Carbon monoxide + Water

HT $2CH_{4(g)} + 3O_{2(g)} \longrightarrow 2CO_{(g)} + 4H_2O_{(g)}$

Incomplete combustion producing carbon monoxide can occur in faulty gas appliances and other heating appliances. This can be dangerous as carbon monoxide is a toxic gas.

If there is *very* little oxygen available when a hydrocarbon burns, carbon is produced. A sooty, yellow flame is an indication of incomplete combustion because it contains carbon.

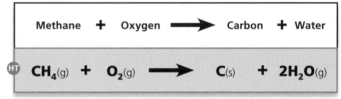

Methane + Oxygen ⟶ Carbon + Water

HT $CH_{4(g)} + O_{2(g)} \longrightarrow C_{(s)} + 2H_2O_{(g)}$

Although hydrocarbons produce useful amounts of energy when they burn, the gases they produce are pollutants.

- Carbon dioxide contributes to an increase in the levels of greenhouse gases, resulting in the greenhouse effect.
- Carbon monoxide, a toxic, colourless and odourless gas, combines irreversibly with the haemoglobin in red blood cells, reducing the oxygen-carrying capacity of the blood. (This eventually results in death through a lack of oxygen reaching body tissues.)

Acid Rain

Hydrocarbon fuels such as petrol, kerosene and diesel oil from crude oil, and methane from natural gas are all **fossil fuels**. Fossil fuels are non-renewable energy sources and can contain impurities such as sulfur. When fossil fuels are burned, **sulfur dioxide** is formed. Sulfur dioxide dissolves easily in rain water to form acid rain.

There are many problems associated with acid rain, which include damage to plants, animals, buildings and metals, and an increase in the acidity levels of many lakes and rivers. As acid rain is formed from gases that dissolve within the moisture in the clouds, it can fall as acid rain many miles away from where the sulfur dioxide was originally released.

Global Warming

Global warming is an accelerated increase in the average temperature of the surface of the Earth.

Much of the heat that reaches us from the Sun is reflected back from the Earth's surface. However, carbon dioxide, methane and water vapour in the atmosphere absorb heat and radiate it back down to Earth. This means that the surface of the Earth is warmer than it would be without these gases.

Scientific evidence shows that the levels of carbon dioxide in the atmosphere have varied over time. In the 20th century, scientists recognised that industrialisation was causing more 'greenhouse gases', including carbon dioxide, to remain inside the Earth's atmosphere (see also page 33).

Throughout the Earth's history, its temperature has fluctuated. Scientists are able to measure the proportion of carbon dioxide in the atmosphere and changes to global temperature over numbers of decades. Using this data, they have found a correlation between the proportion of carbon dioxide in the atmosphere and the global temperature, providing scientists with evidence for climate change.

Chemists have now developed technologies that allow them to remove carbon dioxide from the atmosphere. They can do this by adding iron dust to the ocean which is thought to enhance the growth of plankton, which takes in carbon dioxide (and therefore reduces the amount of carbon in the atmosphere). This is called **iron-seeding**. Chemists can also convert carbon dioxide in the atmosphere into a simple hydrocarbon, by using a catalyst.

Alternatives to Fossil Fuels

Biofuels are fast becoming an attractive alternative to fossil fuels. These are fuels based on sustainable resources such as wood and alcohol from plants.

For some biofuels there is a balance between the carbon dioxide removed from the atmosphere during photosynthesis and the carbon dioxide produced when the fuels are transported and burned.

Ethanol is an alcohol produced by the fermentation of sugar beet and sugar cane. It can be used as a fuel for vehicles in its pure form, but is often added to petrol to reduce the overall need for petrol and to improve vehicle emissions. However, large areas of fertile land are needed to grow the crops.

| Ethanol | + | Oxygen | ⟶ | Carbon dioxide | + | Water |

HT $C_2H_5OH_{(l)} + 3O_{2(g)} \longrightarrow 2CO_{2(g)} + 3H_2O_{(g)}$

A fuel cell is a way of producing and harnessing energy from a chemical reaction. A hydrogen fuel cell uses hydrogen, which reacts with oxygen, usually from the air. **Hydrogen** is the cleanest of all the fuels as it only produces water. If used in cars, it can supply three times the energy of petrol per gram. However, new cars would be needed because of the technology used to get electricity from the hydrogen. Producing the hydrogen involves the use of fossil fuels so still adds to pollution.

The chemical reaction in a simple fuel cell is:

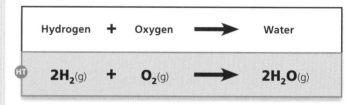

| Hydrogen | + | Oxygen | ⟶ | Water |

HT $2H_{2(g)} + O_{2(g)} \longrightarrow 2H_2O_{(g)}$

Good Fuels

There are many factors to be considered when deciding if a fuel is a good fuel, including these examples.

- Is the fuel safe in that it does not catch alight too easily, especially when transported?
- Is the fuel easy to transport? Liquid fuels are far easier to transport and store, compared to gases.
- Is the fuel a good provider of heat energy so that not too much fuel has to be burned for the energy generated?
- Is it is a clean fuel that produces very little particulate or toxic products, or pollutants that contribute to greenhouse gases?

Good Fuels (cont.)

Different fuels release different amounts of energy when burned. Fuels can be compared by measuring the temperature rise of a fixed volume of water as it is heated by a known amount of fuel.

Alkanes (Saturated Hydrocarbons)

An alkane is a hydrocarbon in which each carbon atom is bonded to four other atoms, each of which is either a carbon or a hydrogen atom. When more than one carbon atom is present they are joined by a **single covalent carbon–carbon bond** to form a chain or 'spine' that runs the length of the molecule. We say that alkanes are **saturated** because each carbon atom is bonded to the maximum number of atoms.

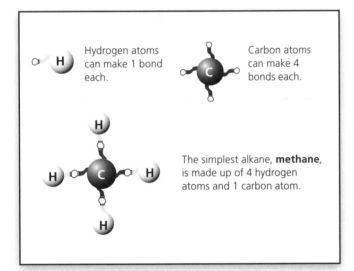

Hydrogen atoms can make 1 bond each.

Carbon atoms can make 4 bonds each.

The simplest alkane, **methane**, is made up of 4 hydrogen atoms and 1 carbon atom.

The next simplest alkanes are **ethane** and **propane**. Because they are saturated (i.e. the carbon atoms are all bonded to four other atoms), they are fairly unreactive, although they do burn well.

Ethane, C_2H_6
A molecule made up of 2 carbon atoms and 6 hydrogen atoms.

Propane, C_3H_8
A molecule made up of 3 carbon atoms and 8 hydrogen atoms.

Alkenes (Unsaturated Hydrocarbons)

The **alkenes** are another kind of hydrocarbon. They are very similar to the alkanes, except that two of the carbon atoms are joined by a **double covalent carbon–carbon bond**. Because the carbon atoms are not bonded to the maximum number of atoms, we call them **unsaturated hydrocarbons**.

The simplest alkene is **ethene**, C_2H_4, which is made of four hydrogen atoms and two carbon atoms. Ethene contains one double carbon–carbon bond.

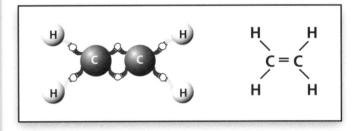

The next simplest alkene is **propene**, C_3H_6, which is made of six hydrogen atoms and three carbon atoms. Propene contains one double carbon–carbon bond and one single carbon–carbon bond.

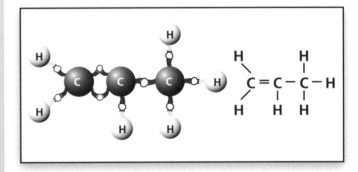

Testing for Alkanes and Alkenes

A simple test to distinguish between alkanes and alkenes is to add bromine water.

Alkenes will **decolourise** bromine water as the alkene reacts with it. Alkanes have no effect on bromine water.

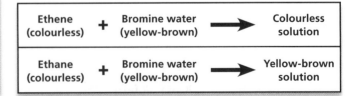

| Ethene (colourless) | + | Bromine water (yellow-brown) | → | Colourless solution |
| Ethane (colourless) | + | Bromine water (yellow-brown) | → | Yellow-brown solution |

Cracking

Cracking is the breaking down of long-chain hydrocarbon molecules into more useful short-chain hydrocarbon molecules, which are then used to help meet demands for hydrocarbons in short supply.

Long-Chain Hydrocarbon

Short-Chain Hydrocarbons

The long-chain hydrocarbon is heated until it vaporises. The vapour is then passed over a heated catalyst where a **thermal decomposition** reaction takes place.

In the laboratory, cracking can be carried out using the following apparatus:

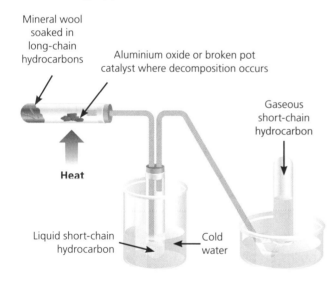

Mineral wool soaked in long-chain hydrocarbons

Aluminium oxide or broken pot catalyst where decomposition occurs

Gaseous short-chain hydrocarbon

Heat

Liquid short-chain hydrocarbon

Cold water

When alkanes are cracked, alkanes and alkenes are formed, for example:

Long-chain hydrocarbon	Heat + catalyst →	Short-chain hydrocarbons
Decane	→	Octane + Ethene
$C_{10}H_{22}$	→	C_8H_{18} + C_2H_4

Supply and Demand

This bar chart shows:

- the relative amounts of each fraction in crude oil
- the demand for each fraction in crude oil.

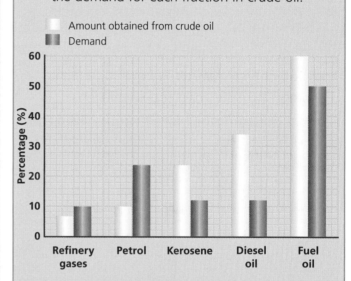

☐ Amount obtained from crude oil
▨ Demand

The demand for some fractions is **greater than** the supply, especially for shorter-chain hydrocarbons, such as petrol. This is because they release energy more quickly by burning, so they make better fuels.

Longer-chain hydrocarbons are broken down into more useful shorter-chain hydrocarbons through cracking. The compounds that are produced have a variety of uses, for example to fuel transport, to heat homes and to make chemicals for drugs.

As crude oil runs out, supply decreases, but demand remains the same and so prices increase.

Scientists are always trying to find alternative ways to make us less reliant on fossil fuels.

Polymers

Long-chain hydrocarbon molecules are often **polymers** made up from smaller **monomers**. A monomer is a short-chain unsaturated hydrocarbon molecule.

Alkenes can be good at joining together because they are unsaturated. When they join together without producing another substance, we call this **addition polymerisation**. For example ethene monomers form the polymer poly(ethene).

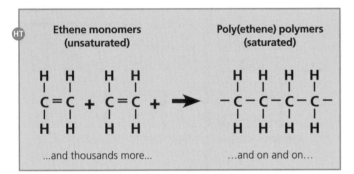

Ethene monomers (unsaturated) → **Poly(ethene) polymers (saturated)**

...and thousands more... ...and on and on...

The general formula for addition polymerisation is:

$$n \left(\begin{array}{c} | \ \ | \\ C = C \\ | \ \ | \end{array} \right) \longrightarrow \left(\begin{array}{c} | \ \ | \\ C - C \\ | \ \ | \end{array} \right)_n$$

where n is a very large number

We can use this formula with different monomer units to show how poly(propene), poly(chloroethene) and PTFE are formed.

Poly(propene):

$$n \left(\begin{array}{cc} H & CH_3 \\ | & | \\ C & = C \\ | & | \\ H & H \end{array} \right) \longrightarrow \left(\begin{array}{cc} H & CH_3 \\ | & | \\ C & - C \\ | & | \\ H & H \end{array} \right)_n$$

Poly(chloroethene):

$$n \left(\begin{array}{cc} H & H \\ | & | \\ C & = C \\ | & | \\ H & Cl \end{array} \right) \rightarrow \left(\begin{array}{cc} H & H \\ | & | \\ C & - C \\ | & | \\ H & Cl \end{array} \right)_n$$

PTFE:

$$n \left(\begin{array}{cc} F & F \\ | & | \\ C & = C \\ | & | \\ F & F \end{array} \right) \rightarrow \left(\begin{array}{cc} F & F \\ | & | \\ C & - C \\ | & | \\ F & F \end{array} \right)_n$$

Uses of Some Polymers

Generally polymers such as poly(ethene), poly(propene), poly(chloroethene) (PVC) and PTFE are very versatile as they have a variety of different uses.

Polymer	Uses
Poly(ethene), **PE** Good chemical resistance, flexible	Food coverings, bins, water pipes, bags, bottles, food trays
Poly(propene), **PP** Good chemical resistance, flexible	Chairs, stationery, ropes, crates, bottles, clothing
Poly(chloroethene), **PVC** Good impact resistance, rigid or flexible	Flooring, double-glazed window frames, pipes, clothing, cable covering, electrical components
PTFE Water-resistant, flexible, non-stick	Non-stick cookware coating, lubricant sprays, to make waterproof coatings

Disposing of Polymers

There are various ways of disposing of polymers. Unfortunately, some of these methods can be harmful to the environment.

- **Burning polymers** produces air pollution. Some polymers should not be burned as they produce toxic fumes, for example burning PVC produces hydrogen chloride gas.
- **Use of landfill sites** means that plastic waste builds up because most polymers are non-biodegradable. Microorganisms have no effect on them: they will not decompose and rot away.
- **Recycling** some polymers uses less energy to make a new product, reduces the need for oil in polymer production, reduces the amount of polymer waste going to landfill and helps reduce carbon dioxide emissions.

More and more companies have made investments into biodegradable polymers, which are continually being developed.

Questions labelled with an asterisk (*) are ones where the quality of your written communication will be assessed – you should take particular care with your spelling, punctuation and grammar, as well as the clarity of expression, on these questions.

1 (a) An acid–base reaction is called:

A ☐ oxidation **B** ☐ neutralisation **C** ☐ reduction **D** ☐ precipitation **(1)**

(b) What type of product would be formed in this type of reaction? **(1)**

2 (a) How is the atmosphere today different from the Earth's early atmosphere? **(2)**

(b) Describe how water vapour from early volcanic activity formed the oceans. **(2)**

3 State three ways in which the levels of carbon dioxide in the Earth's atmosphere have increased over the last 100 years. **(3)**

4 (a) What is a hydrocarbon? **(1)**

(b) State one way in which increasing the number of carbon atoms in a hydrocarbon affects its properties. **(1)**

(c) Why can crude oil be considered as a mixture? **(1)**

(d) Two hydrocarbons obtained from a cracking process are ethene and decane. A sample of each is added to bromine water. What would you expect to happen when ethene is added to the bromine water?

A ☐ The solution turns yellow.

B ☐ The solution turns from orange to colourless.

C ☐ The solution turns milky.

D ☐ The solution remains orange. **(1)**

5 (a) Give one economic advantage to recycling metals. **(1)**

*****(b)** Explain in as much detail as you can why recycling can be considered as sustainable. **(6)**

6 (a) What is an ore? **(1)**

(b) Which property of a metal determines how easily it is extracted from an ore? **(1)**

(c) (i) Copper oxide can be heated with carbon to extract the metal. Complete the following equation.

Copper oxide + Carbon ⟶ .... + ... **(1)**

(ii) What has happened to the copper oxide in the above equation? **(1)**

(d) What does the term 'reduction' mean? **(1)**

7 Explain why the method used to extract aluminium from its oxide is different from that used to extract iron from its oxide. **(4)**

8 (a) What is produced when calcium carbonate is decomposed? **(2)**

(b) Limestone is a very important building material that can be broken down by thermal decomposition. Why is limestone important in the building industry? **(2)**

(c) Describe four factors that should be taken into consideration when quarrying for limestone. Your answer must include at least one each of the environmental, economic and social effects. **(4)**

9 Acid and alkali reactions all follow the same general equation.

(a) Write the word equation for the reaction of sulfuric acid and magnesium oxide. **(1)**

(b) Write the word equation for the reaction of nitric acid and copper oxide. **(1)**

(c) Describe why ingesting indigestion remedies is the same as a neutralisation reaction. **(2)**

(d) Why does the stomach contain hydrochloric acid? **(2)**

10 (a) Describe the difference between a monomer and a polymer. **(2)**

(b) Give the structural formula for PTFE. **(1)**

(c) Choose the correct answer to complete the following sentence.

When monomers join together without producing another substance, it is called

A ☐ co-polymerisation.   B ☐ addition polymerisation.

C ☐ polymerisation.   D ☐ condensation polymerisation. **(1)**

(d) Explain why burning polymers is not an environmentally friendly way of disposing of them. **(3)**

(e) State one property of PTFE that makes it useful as a lubricant. **(1)**

11 Chlorine is a very important element. It can be obtained from the electrolysis of sea water and from brine, which is the basis of the chlor-alkali industry.

(a) Describe how electrolysis breaks down brine to release chlorine gas and hydrogen gas. **(3)**

(b) (i) Describe the appearance of chlorine gas. **(1)**

(ii) How would you test for chlorine gas at the positive terminal of an electrolysis cell? **(1)**

(c) Why is the large-scale manufacture of chlorine considered to be potentially hazardous? **(2)**

12 *Iron is a very versatile metal. It is extracted from haematite, iron(III) oxide, by reacting it with carbon monoxide in a blast furnace. Pure iron has very few uses as it is too soft. To make it more useful iron is mixed with other elements.

Explain in as much detail as you can how mixing iron with other elements improves its properties. **(6)**

HT

13 This question is about fuels.

(a) (i) Methane has the formula CH_4 and is the fuel that is often called 'natural gas'. It is also an alkane. Write a balanced equation for complete combustion of methane in oxygen. **(1)**

(ii) When methane reacts with insufficient oxygen, incomplete combustion takes place. Write a word equation and a balanced symbol equation for this reaction. **(2)**

*(b) Petrol and diesel oil are fractions of crude oil. As a fuel, petrol is in greater demand (25% of demand for fuels) than can be supplied (10% obtained from crude oil). This is not the case for diesel oil (12% of demand for fuels, but 34% obtained from crude oil).

Describe as fully as you can what you understand by the supply and demand of crude oil, using the examples given, and how crude oil refinery overcomes this problem. **(6)**

This topic looks at:
- how ideas about the Solar System have changed
- how visible light is used to make observations
- how a reflecting telescope works

Changing Ideas about the Solar System

In the 6th century BC, the Ancient Greeks believed that the Earth was at the centre of the Solar System. This theory is known as the **geocentric model**.

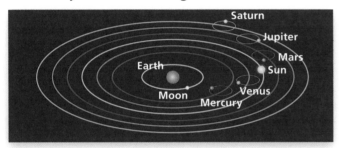

The theory of the geocentric model was supported by the observations at that time. For example, it was generally accepted that the Earth could not be moving or the motion would have been felt. It was also believed that the five visible planets, **Mercury**, **Venus**, **Mars**, **Jupiter** and **Saturn**, were attached to crystalline spheres. These spheres were set one inside another and revolved around the Earth.

It was not until the 16th century that a sun-centred Solar System (also known as a **heliocentric model** from 'helios', the Greek word for Sun) was seriously considered. This idea was put forward by the Polish astronomer, Copernicus. The following observations contributed to the evidence for the heliocentric model.

In 1610, the Italian physicist, Galileo, became the first person to make telescopic observations of the planets Venus and Jupiter. He observed that the phases of Venus were similar to the phases of the Moon. This went against the geocentric model, since, according to this model, Venus should always have appeared as a crescent shape.

Galileo also observed that Jupiter had four orbiting moons. (He could only see the four largest of the 63 confirmed moons that Jupiter actually has.) According to the geocentric model, everything orbited around the Earth.

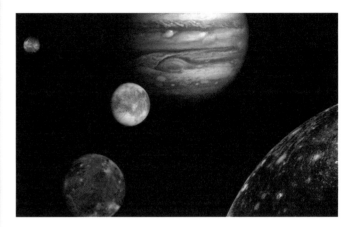

Before the heliocentric model, it was also believed that the planets took circular paths, as circles were considered to be perfect shapes. However, at the end of the 16th century, Johannes Kepler, a German mathematician, discovered that the orbits are in fact ellipses (squashed circles).

It was not until some time after Galileo's death, though, that the heliocentric model of the Solar System became generally accepted.

The Discovery of Uranus, Neptune and Pluto
Uranus was not officially discovered until 1781. It was viewed through a telescope (though it can just be seen with the naked eye).

Neptune, which lies further away, was more difficult to discover. Neptune was discovered in 1846 because it was observed that **Uranus** appeared to be out of its calculated position in the sky. It was concluded that another planet was pulling on it. This planet was Neptune.

Due to its immense distance from Earth, **Pluto** (once thought of as the ninth planet but now down-graded to a dwarf planet) was not discovered until 1930. The night sky was repeatedly photographed, in searches for movement against the background of stars. However, as Pluto appeared so tiny in the photographs, it was easy to miss.

Waves

Waves are regular patterns of disturbance. They transfer energy and information from one point to another without any transfer of matter. Waves can be produced in ropes, springs and on the surface of water.

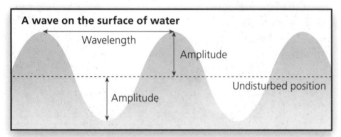

A wave on the surface of water
- Wavelength
- Amplitude
- Amplitude
- Undisturbed position

- **Amplitude** is the maximum vertical disturbance caused by a wave (i.e. its height).
- **Wavelength** is the distance between corresponding points on two successive disturbances.
- **Frequency** is the number of waves produced (or passing a particular point) in one second.

There are two types of wave, which can be demonstrated using a slinky spring.

1 Transverse waves – the pattern of disturbance is at right angles (90°) to the direction of wave movement.

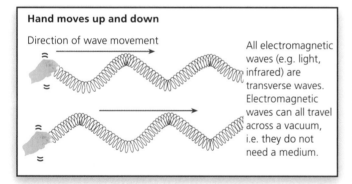

Hand moves up and down

Direction of wave movement

All electromagnetic waves (e.g. light, infrared) are transverse waves. Electromagnetic waves can all travel across a vacuum, i.e. they do not need a medium.

2 Longitudinal waves – the pattern of disturbance is in the same direction as the direction of wave movement.

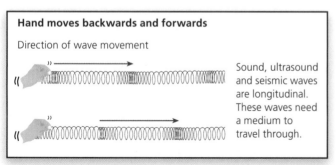

Hand moves backwards and forwards

Direction of wave movement

Sound, ultrasound and seismic waves are longitudinal. These waves need a medium to travel through.

Similarities
- Both types of wave carry energy.

Differences
- They travel at different speeds.
- The vibrations (patterns of disturbance) are different.
- Longitudinal waves need a medium to travel through; some transverse waves (e.g. electromagnetic spectrum) do not.

Refraction and Reflection

When a ray of visible or infrared light travels from glass, Perspex or water into air, it is **refracted** (changes direction).

Some light is also **reflected** from the boundary. The angle of incidence = the angle of reflection.

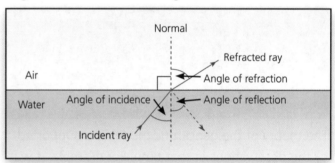

Normal
- Air
- Refracted ray
- Angle of refraction
- Water
- Angle of incidence
- Angle of reflection
- Incident ray

HT This refraction is **away** from the normal. (The normal is the line at right angles to the boundary at the point of incidence.) The light **speeds up** as it passes into the air.

When a ray travels from air into glass, Perspex or water, it is also refracted.

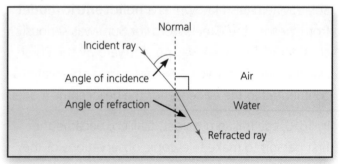

Normal
- Incident ray
- Angle of incidence
- Air
- Angle of refraction
- Water
- Refracted ray

HT In this case, refraction is **towards** the normal because the light **slows down** when it enters the more dense glass, Perspex or water.

Wave Speed

Wave speed, frequency and wavelength are related by the equation:

$$\text{Wave speed (m/s)} = \text{Frequency (Hz)} \times \text{Wavelength (m)}$$

$$\frac{v}{f \times \lambda}$$

where λ is the wavelength, f is the frequency and v is the wave speed

Example 1

A sound wave has a frequency of 168Hz and a wavelength of 2m. What is the speed of sound?

Wave speed = Frequency × Wavelength

$$= 168 \times 2$$

$$= \textbf{336m/s}$$

Example 2

Radio 5 Live transmits on a frequency of 909kHz. If the speed of radio waves is 300 000 000m/s, what is the wavelength of the waves?

$$\textbf{Wavelength} = \frac{\textbf{Wave speed}}{\textbf{Frequency}}$$

$$= \frac{300\,000\,000}{909\,000} \quad \leftarrow \text{Change kHz to Hz}$$

$$= \textbf{330m}$$

Wave speed can also be found by measuring the distance travelled by the wave in a certain time. So we can use the equation:

$$\text{Wave speed (m/s)} = \frac{\text{Distance travelled (m)}}{\text{Time taken (s)}}$$

$$\frac{x}{v \times t}$$

where x is the distance travelled

Example 3

The average time for a water wave to move a distance of 10m is estimated to be 4 seconds. What is the speed of the waves?

$$\textbf{Wave speed} = \frac{\textbf{Distance}}{\textbf{Time}}$$

$$= \frac{10}{4}$$

$$= \textbf{2.5m/s}$$

Observing the Universe

The Universe emits **radiation** over the entire electromagnetic spectrum of waves, including light waves.

There are different ways of using visible light to make observations.

Before Galileo's telescopic observations in 1610, all observations were made by the **unaided** or **naked eye**. Approximately 2000 stars can be seen with the naked eye, as well as the Moon and Uranus. However, what can be seen with the naked eye depends on weather conditions and also on the eyesight of the observer. In cities, light pollution can pose a problem for observing the night sky.

Photographing the night sky allows areas of it to be recorded for examination later. But since light levels are low at night, the camera shutter must be open for a long time to let enough light in. Therefore, due to the rotation of the Earth, stars and other bodies will appear as streaks rather than pin-points of light. The camera must also be mounted on a tripod so there is no movement during the long exposure.

With a small **telescope** (again depending on viewing conditions), the following can be seen:
- the phases of Venus
- the four so-called Galilean moons of Jupiter
- the rings of Saturn
- Uranus
- Neptune.

A telescope with a larger diameter lens will allow more detail and many more stars to be seen. For example, a 20cm diameter reflecting telescope will allow perhaps a million stars to be viewed.

All of the stars in the night sky are part of our galaxy, called the Milky Way. Our Sun orbits the centre of the galaxy, which can sometimes be seen as a bright band stretching across the sky.

Making a Telescope

A simple telescope consists of two different **converging lenses**. A converging lens is made of Perspex or glass with two curved surfaces, thicker in the middle than at the edges. This shape allows the lens to bring together (converge) light to a focus by refracting it.

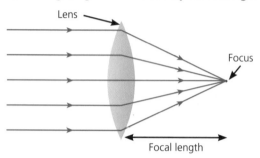

The distance from the middle of the lens to the focus is called the **focal length**. If the lens is held up to the light (e.g. a window), an upside-down image (of the window) can be seen on a piece of paper held on the other side. An image that can be projected onto a screen is said to be a **real image**.

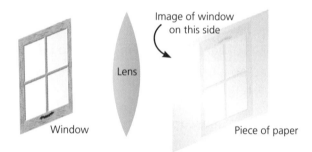

The piece of paper or the lens can be moved until the image is sharp (focused). The distance from the lens to the image is the focal length, which can be measured. For a given type of glass, the thickness of the lens will determine its focal length – so a thicker lens means a shorter focal length.

If a converging lens is held near to an object (e.g. the print in this book) an upright, magnified image will be seen. This image cannot be projected onto a screen. It is said to be a **virtual image**.

To make a telescope, you need two lenses – a thin one and a smaller, thick one. Hold the thick one near the eye. This lens is called the eyepiece. It magnifies the image produced by the other lens. The thin lens is called the objective and collects light from the object.

Magnification depends on the focal lengths of the lenses. Experiment with magnification by using converging lenses of different thicknesses:

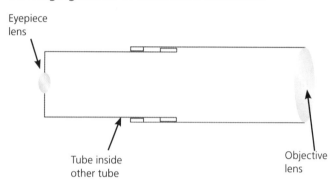

1. Hold the eyepiece up to the eye with one hand.
2. Hold the objective with the other hand outstretched.
3. Look at the view out of a window and move the objective lens until a clear image is seen. Two cardboard tubes to mount the lenses will make it easier to use.

However, this type of simple telescope is limited because it is expensive to make large, good quality lenses. This is why, when Isaac Newton built the first successful **reflecting telescope** in 1668, he used two mirrors and an eyepiece.

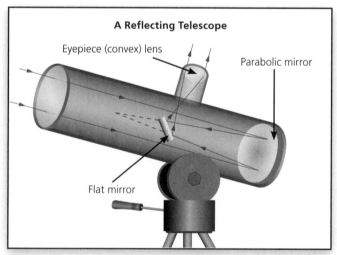

Light from a distant object strikes the large parabolic mirror at the end. It is then reflected to a small secondary flat mirror. Finally, the light is reflected onto an eyepiece through which the image is seen.

Reflectors gather much more light (and so see more) than refractors do, because it is easier to make large mirrors than large lenses.

P1 Topic 2: The Electromagnetic Spectrum

This topic looks at:

- the order of the waves in wavelength or frequency
- the uses of different types of wave
- the dangers of waves

The Discovery of Waves Outside the Visible Spectrum

In 1666, Isaac Newton discovered the **visible spectrum**. The visible spectrum is the range of light waves that can be seen by the human eye.

It was not until 1800 that **William Herschel** discovered infrared (heat) waves. Herschel used a row of thermometers to measure the temperature of different parts of the visible spectrum. He noticed that the temperature increased slightly as he moved the thermometers to the red end of the visible spectrum. When he then moved them into the dark region beyond the red end, Herschel was surprised to note that the temperature increased rapidly.

In 1801, **Johann Ritter** was interested in the chemical reaction of silver chloride. Silver chloride breaks down to black silver when exposed to light (which is the basis of photographic film). When Ritter moved the silver chloride to the violet end of the spectrum, it reacted a little faster. When moved into the dark region beyond the violet, it reacted very quickly. Ritter had discovered ultraviolet rays.

In 1888, **Heinrich Hertz** sent a signal across his laboratory and so discovered radio waves.

In 1895 **Wilhelm Röntgen** discovered X-rays, and in 1900 **Paul Villard** discovered gamma rays.

Electromagnetic Waves

Energy from the Sun travels to the Earth in the form of **electromagnetic waves**. These waves form the **electromagnetic spectrum** in which they are ordered according to their frequency and wavelength. All these waves carry energy, are transverse and travel at the speed of light (in a vacuum).

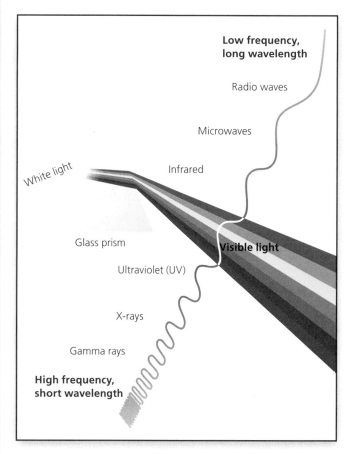

The electromagnetic spectrum is continuous from radio to gamma rays. However, as can be seen from the above diagram, the waves can be grouped in order of **decreasing wavelength** and **increasing frequency**. That is, from radio waves to gamma rays. One way to remember this order is: "**R**eading **M**agazines **I**s **L**ousy **U**nless e**X**tremely **G**ood".

The colours of the visible spectrum can be remembered by the phrase **ROY G BIV** - **r**ed, **o**range, **y**ellow, **g**reen, **b**lue, **i**ndigo, **v**iolet.

Uses of Electromagnetic Waves

Electromagnetic Waves	Uses
Radio waves	Transmitting television and radio programmes, communications between different places and satellite transmissions.
Microwaves	Satellite communication, mobile phones, cooking.
Infrared	Grills, toasters, heaters, remote controls (short-range communication), optical fibre communication, treatment of muscular problems, night vision (thermal imaging) and security systems.
Visible light	Vision, photography, illumination.
Ultraviolet	Fluorescent lamp and security coding, sunbeds, detecting forged banknotes and disinfecting water.
X-rays	Producing shadow pictures of bones and metals to observe the internal structure of objects and animals in medical X-rays, airport security scanners.
Gamma rays	Detecting cancer and treating it by killing cancer cells, killing bacteria on food and sterilising surgical instruments.

Absorption and Emission

Microwaves to Monitor Rain

Microwaves have a wavelength suitable for absorption by water molecules. Satellites are able to monitor how the microwaves are absorbed by the atmosphere, showing areas of probable high rainfall.

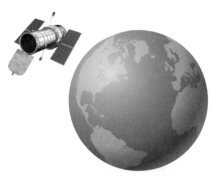

Infrared Sensors

Infrared sensors can detect temperature differences of surfaces because the higher the temperature, the more infrared radiation is emitted. Police helicopters use infrared sensors to follow suspects at night, or when the suspect hides somewhere such as in woodland. Rescuers can detect infrared radiation from people trapped in collapsed buildings, and some alarm sensors use infrared to detect movements.

X-rays to See Bone Fractures

The area with the suspected fracture is placed in front of a photographic plate and is exposed to X-rays. The X-rays are absorbed by the bone. However, they pass through the fracture and expose the photographic plate, clearly showing where the fracture is.

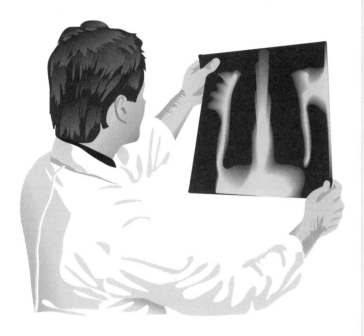

Ultraviolet to Detect Forged Bank Notes

If paper is exposed to ultraviolet, some is absorbed. The paper and inks then emit visible light (fluorescence). Different papers and inks fluoresce differently, which can help in detecting forgeries. Security marker-pens, which glow under ultraviolet light, can be used to mark property.

Dangers of Waves

The higher the frequency of the waves, the more damage they can cause with excessive exposure. Gamma rays have the highest frequency, and radio waves have the lowest frequency.

Microwaves can be absorbed by the water in the cells in our body. This can cause internal heating of body tissue, which may damage or kill cells. They also make magnetic fields, which can affect how the cells work.

Infrared is absorbed by the skin and felt as heat. Too much exposure will cause burns.

Ultraviolet (**UV**) passes through the skin into the tissues. Darker skin allows less penetration and therefore offers more protection. The greater the amplitude the more energy the wave will carry and the more likely it is to be dangerous. Low amplitude UV is absorbed by the Earth's atmosphere.

Short exposure to UV can kill normal cells. Longer exposure can result in skin cancer. UV exposure also means you are more likely to develop cataracts in the lens of the eye.

There are three types of UV radiation:

Increasing frequency

1. **UVA** – passes through glass, penetrates deep into the skin, causes early ageing, wrinkles, DNA damage, cancer and some sunburn.
2. **UVB** – mostly absorbed by the ozone layer and the atmosphere. Dangers are the same as for UVA but UVB also stimulates production of essential vitamin D.
3. **UVC** – almost all is absorbed by the ozone layer and the atmosphere. Most damaging.

We are at most risk from UVB, as almost all UVC is absorbed by the ozone layer.

X-rays and **gamma rays** pass through soft tissues (although some rays are absorbed). Short exposure can kill normal cells. Longer exposure can result in cancer, through the destruction or mutation of cells. Gamma rays ionise material that they pass through, transferring energy and damaging living cells.

Stars, Galaxies and the Universe

Our Sun is one of many billions of stars in the **Milky Way**. A collection of many stars is called a galaxy. The Milky Way is just one galaxy of many millions of **galaxies** in the **Universe**.

Not to scale

Our Sun

Our Sun —

Our galaxy the Milky Way

Our galaxy

The Universe

The Solar System

The **Solar System** is made up of the **Sun** (a star) and the eight **planets** (plus Pluto, which is now classed as a dwarf planet) that surround it. These planets move around the Sun in paths called **orbits**, which are slightly elliptical (squashed circles).

Planet	Diameter (km)	Distance from Sun (million km)
Mercury	4 880	58
Venus	12 112	107.5
Earth	12 742	149.6
Mars	6 790	228
Jupiter	142 600	778
Saturn	120 200	1 427
Uranus	51 000	2 870
Neptune	49 200	4 497
Pluto (dwarf planet)	2 284	5 900

The table of data gives the diameters of the planets in the Solar System, and their average distances from the Sun. The Moon, at about 380 000km away, is closest to the Earth. It took the Apollo astronauts less than four days to get there. The Moon's diameter is roughly a quarter that of the Earth, which means that about 50 moons could fit inside the volume of the Earth.

Jupiter's diameter of about 143 000km is over 11 times the diameter of the Earth. Even Jupiter is dwarfed by the Sun though, which has a diameter of nearly 1 400 000km. The Sun's volume is over 1 million times the volume of the Earth.

The Sun appears to be much bigger than other stars in the sky because it is much nearer to Earth, but it is a small-to-average sized star. The Sun and Moon actually appear the same size because, although the Sun is 400 times larger in diameter, it is also 400 times further away.

Beyond the Solar System

The nearest star outside the Solar System is one named Proxima Centauri. It takes about 4.3 years for its light (travelling at 300 million metres per second) to reach the Earth. Compare that to the 8.3 minutes for the light from the Sun to reach Earth.

Distances to further bodies are so enormous that astronomers measure distances in **light years**. A light year is the distance that light travels in one year.

Using Modern Telescopes

Much data is gathered from telescopes that use all parts of the electromagnetic spectrum as well as visible light.

Stars, galaxies and other bodies emit large amounts of energy over the whole of the electromagnetic spectrum. Astronomers use **radio**, **infrared**, **ultraviolet**, **X-ray** and **gamma ray** telescopes to observe these invisible emissions from the Universe.

Radio Telescopes

Radio telescopes, like telescopes that use visible light and infrared, will work from Earth. However, ultraviolet, X-ray and gamma ray telescopes need to be above the Earth's atmosphere in order to work. This is because these radiations are largely absorbed before reaching the Earth's surface.

> **HT** The Earth's atmosphere blocks 98.7% of the UV radiation from penetrating through it – the radiation is absorbed by the ozone layer.
>
> Gamma rays and X-rays, which are shorter in wavelength, are absorbed by oxygen and nitrogen.

> **HT** UV, X-rays and gamma rays are selectively scattered much more than longer wavelengths. This scattering is caused by gas molecules, smoke fumes, etc. The scattering is broadly inversely proportional to the wavelength of the radiation.
>
> Even telescopes that detect visible light can take much sharper images if they are situated in orbiting satellites outside the atmosphere. Atmospheric distortion, light scattering, light pollution and general problems with weather all limit ground-based telescopes.

Data Gathered by Modern Telescopes

The **Hubble Space Telescope** was launched by NASA in April 1990. It orbits the Earth outside the atmosphere, and has taken some spectacular images of galaxies by visible light.

The Hubble Space Telescope has contributed enormously to our understanding of the Universe. The Hubble Deep Field and Ultra Deep Field images have been constructed using higher and higher magnifications of the Universe. These images have provided evidence to show how galaxies have evolved (they tend to start off irregular in shape and gradually become elliptical). The images have also led to the discovery of more and more galaxies, up to billions of light years away.

Data Gathered by Modern Telescopes (cont.)

A **radio telescope** set up at Cambridge University in 1968 discovered a strange pulsing radio signal from outer space. At first, it was thought that this might be from intelligent life. Now the signal is explained as being from an extremely dense rotating star (a neutron star) called a **pulsar**. This sends out a signal much like the light from a lighthouse.

In 2009, astronomers using **NASA's Swift satellite** were the first to record a gamma ray burst from a collapsing star, spewing out jets of gas. It was confirmed to be the most distant object ever observed in the Universe. It is a staggering 13.1 billion light years from the Earth.

The **Planck observatory satellite** has collected an enormous amount of data, mapping the whole sky and providing a good estimate for the age of the Universe and the rate of its expansion.

Searching for Intelligent Life

There are two ways of looking for intelligent life in the Universe:

1. Sending spaceships out to collect and return data. The main problems with this are the enormous distances and journey times involved – thousands of years.
2. Searching for radio signals. Radio telescopes receive information all the time. They are more useful than light-collecting telescopes when looking for alien signals as light can be blocked by dust particles and gas.

The position of a planet within its solar system determines its potential for the existence of life. A planet should be within a 'habitable zone' orbiting its star (i.e. a similar distance to that of the Earth from the Sun).

It is unlikely that our two nearest neighbours (Venus and Mars) have any life. For example, Venus is too hot (its dense atmosphere gives a huge greenhouse effect) and its atmosphere would crush us.

Between 1990 and 2005, 130 stars with orbiting planets were found, and the first image was produced in 2004. Scientists used other evidence for the existence of planets, such as changes in brightness of stars as orbiting planets obscured their light, and a 'wobble' in the stars' motion caused by the planets' gravity. Planets around stars outside our solar system are called **exoplanets**.

In 1992, NASA set up **SETI** (Search for Extraterrestrial Intelligence), which looks for radio signals that may have been emitted by aliens. Using SETI@home, over 50 000 people around the world are helping to process data.

Unmanned Space Exploration

The distances involved in exploring the Solar System, let alone exploring the Milky Way, are huge. It would take several years for a spacecraft to travel to Pluto. It is not realistic to send manned spacecraft on such long journeys, so data logging and remote sensing are required, whereby information can be sent to receivers on Earth via radio waves.

Unmanned craft are often used in space exploration because:
- they are safer
- the journey is so long
- the equipment is as effective (or more so) as humans (e.g. collecting soil / rock from a planet or moon surface and performing an analysis).

Examples of unmanned spacecraft:
- **Viking Lander** (1975) took images of the surface of Mars and analysed the atmosphere and soil.
- **NASA Spirit and Opportunity Rovers** are currently investigating Mars.

NASA's Mars Exploration Rover

The Life and Death of a Star

Stars may be at different stages in their life cycles. They do not last forever, and some stars we can see no longer exist; it takes 4.3 years for the light from the closest star to reach Earth and thousands of years for that from the more distant stars. Our Sun is a small-to-average sized star.

Star Formation

Stars are made from **nebulae** (clouds of gases and dust) that are pulled together, or collapsed, by gravitational forces. This increases the temperature and nuclear reactions start to take place, releasing massive amounts of energy and forming a star.

When stars fuse hydrogen into helium, generating light and heat, they are in a stable state known as the **main sequence**. Hydrogen is abundant in stars, so most stay on the main sequence for a long time, giving life the chance to evolve.

Star Formation

Nebula

Star forms

Star and planets

Death of a Star

Eventually, the required hydrogen gas runs out, causing the star to expand and get colder. What happens next depends upon the size of the star.

A star the size of the Sun becomes a **red giant**. It continues to cool before collapsing under its own gravity to become a **white dwarf**, and then finally a **black dwarf**:

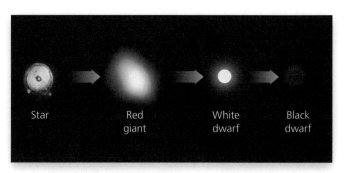

Star | Red giant | White dwarf | Black dwarf

HT A star much bigger than the Sun becomes a **red supergiant**. It shrinks rapidly and explodes, releasing massive amounts of energy, dust and gas into space. This is a **supernova**. The dust and gas (nebula) will form new stars and the remains of the supernova will become either a **neutron star** or a **black hole**:

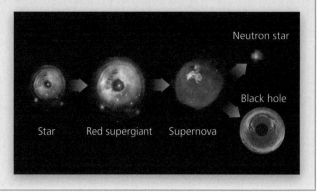

Neutron star

Black hole

Star | Red supergiant | Supernova

Theories of the Origin of the Universe

Steady State Theory: this theory assumes that the Universe does not change its appearance over time. The Universe has no beginning and no end, according to this theory.

Big Bang Theory: this theory assumes that the Universe started about 15 billion years ago when a massively dense object experienced a tremendous explosion known as the Big Bang. Since then, according to this theory, the Universe has been continually expanding.

Big Bang Theory (cont.)

The diagram shows the expansion of the Universe after the Big Bang.

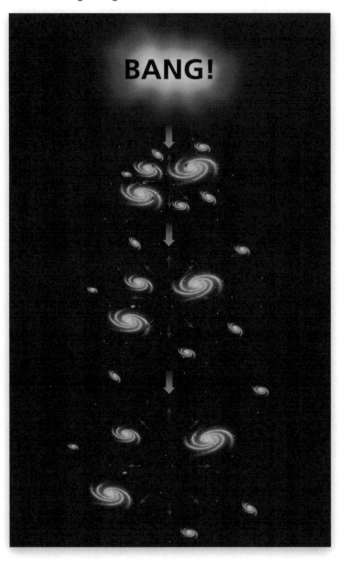

Evidence for the Big Bang Theory

Red-shift

If a source of light waves, such as a galaxy, is moving away from or towards us, the frequency and wavelength of the light that we see will change. It looks different compared with light from a source that is not moving in relation to us.

Studies of light from distant galaxies show that they are moving away from us (i.e. the Universe is expanding). Light from these galaxies is 'shifted' towards the red part of the visible spectrum; this is known as **red-shift**.

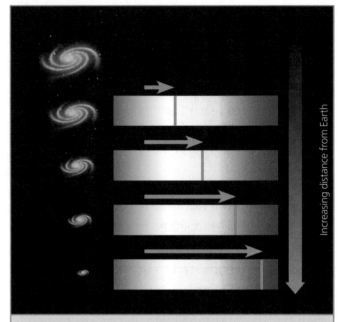

Increasing distance from Earth

HT The blue lines in the diagram show where some light is absorbed by elements in the atmosphere around stars in a galaxy. The more distant the galaxy, the more this absorption line is shifted towards the red end of the spectrum. This shows that the Universe is expanding, and that the more distant the galaxy, the faster it is moving away from us.

The Steady State Theory also says the Universe is expanding, but that new matter is continually created to fill the space.

Cosmic Microwaves

Cosmic microwaves have been detected. This **cosmic microwave background** (**CMB**) radiation shows that the Universe is cooling (i.e. it started very hot and is cooling as it expands).

HT The CMB is spread uniformly through space and was predicted by the Big Bang Theory as radiation left over from the moment of creation. The Steady State Theory has tried but failed to explain this.

Overall, there appears more evidence to support the Big Bang Theory and this is the currently accepted model.

Explaining the Red-shift

A stationary police car puts on its siren. The sound waves reach an observer and a sound is heard. The sound has a certain frequency that depends on the wavelength.

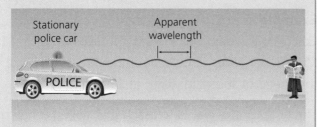

The car now moves off towards the observer. The sound waves now appear to be more bunched together and the wavelength is shorter. So the frequency of the sound heard is higher.

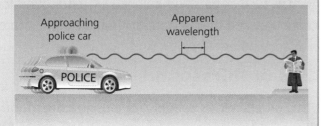

When the car moves away from the observer, the sound waves appear to be pulled further apart, i.e. the wavelength is longer. The frequency of the sound heard is lower.

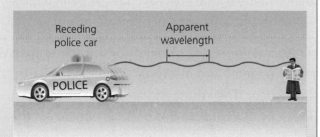

This is true for all waves where the wave source is moving. Red light has a lower frequency than blue light. The light observed from distant galaxies is shifted towards the red, that is, the frequency appears lower. So the galaxies are moving away from us.

Looking at Light Sources

A simple investigation of the light spectra given out by different light sources can be carried out by using an unwanted CD or DVD, and a cardboard box such as a shoe box or cereal box.

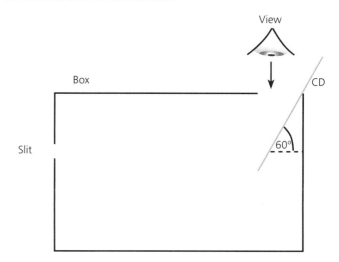

A CD has on its surface a great many very fine grooves. The grooves act as a grating that breaks up light, like a prism does, into its separate colours.

1 The CD should be fixed at an angle of 60° and a narrow slit of about 2mm should be made opposite the lower half of the CD. The slit needs to be carefully made, using thick paper and duct tape, for instance, or with two straight edges such as used razor blades. If the slit is too wide, the spectrum will look blurred; if it is too narrow the spectrum will be too dim to see clearly. Black or duct tape needs to be used to stop stray light getting into the box around the sides of the CD.

2 The slit is then held up to different light sources. For example:
- an overhead fluorescent light
- a reading light
- a candle
- the light from a computer screen.

The spectra should be viewed by looking directly down at the CD.

P1 Topic 4: Waves and the Earth

This topic looks at:
- how ultrasound and infrasound are used
- how to detect earthquakes
- how seismic waves are used to explain the interior of the Earth

Ultrasound

Sound is produced when something vibrates backwards and forwards. Sound waves are longitudinal, travel at the speed of sound, need a medium to carry them and cannot pass through a vacuum. The speed at which a sound wave travels depends on the medium through which it is passing.

A sound can be heard if it is within the audible range of our ears. Most humans can hear sounds in the range of 20 to 20 000 hertz (Hz), which means 20 to 20 000 vibrations per second.

Ultrasound waves have frequencies greater than 20 000Hz.

Infrasound waves haves frequencies less than 20Hz.

Uses of Ultrasound

Ultrasonic waves are used in medicine to produce visual images of different parts of the body (e.g. the heart and liver) to detect problems.

Ultrasound waves are used for foetal scanning in pregnant women, to determine size and position of the foetus, and to detect any abnormalities. It is safe, with no risk to patient or baby (unlike X-rays).

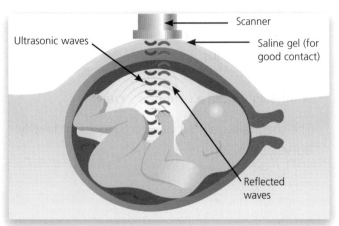

Ultrasound is also used in **sonar**, where waves are sent out from the bottom of a ship.

The reflected waves are received and the time delay of the reflections is used to calculate the depth of the sea at that point or, for example, the depth of a shoal of fish.

Example

Calculate the depth of the sea if it takes 3s for ultrasound waves to be sent out and received. Take the speed of sound in water as 1500m/s.

The time for the waves to reach the seabed is half of 3s (as in this time they must travel there and back).

Depth = Speed × Time
$$= 1500 \times 1.5$$
$$= \textbf{2250m}$$

Animals such as bats and dolphins use ultrasound to locate prey and their surroundings and communicate with each other.

Uses of Infrasound

Many other animals communicate using **infrasound**. These include large animals such as elephants, whales and rhinoceros. African elephants are able to communicate across distances up to 10km at frequencies between 15 and 35Hz. Tigers use infrasound to warn off rivals. It is possible that the sounds made by the animals could be detected by researchers to locate them in remote areas.

Infrasound signals are given off by volcanic eruptions and by meteors and meteorites entering the Earth's atmosphere. The infrasound array built at the University of California has been able to detect the explosions caused by meteors. One of the aims of the Acoustic Surveillance for Hazardous Eruptions project, located in Ecuador and the USA, is to develop a better understanding of violent volcanic eruptions by monitoring the infrasound remotely hundreds of kilometres away.

Earthquakes and Tsunami

The Earth's surface is split into several large **tectonic plates**, which are moving slowly (a few centimetres per year). This movement is due to **convection currents** in the Earth's mantle. It is the movement of these plates that causes **earthquakes**.

When plates slide past each other, the movement is not smooth and the plates get stuck. This causes pressure to build up and earthquakes occur when the pressure is released.

Tectonic Plates

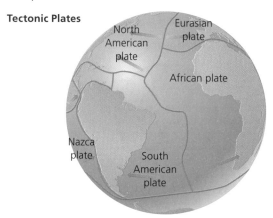

A **tsunami** is caused by an underwater disturbance, normally an earthquake or volcano. The wave travels fast, has a long wavelength and small amplitude, and stores vast amounts of energy. As it approaches land, the height of the wave increases drastically and it transfers the energy to everything in its way.

It is very difficult to predict when earthquakes and, therefore, tsunamis will occur. Scientists have been trying for hundreds of years. They can predict *where* they will happen, as they know where the faults in the Earth's crust are, but not *when* they will happen. This is because the Earth's tectonic plates do not move in regular patterns. Scientists can measure the strain on underground rocks to evaluate the likelihood of a forthcoming earthquake, but they cannot predict an exact time.

Although we are not able to predict individual earthquakes, the world's largest earthquakes do have a spatial pattern, and estimates of the locations and magnitudes of some future large earthquakes can now be made.

Seismic Waves and the Structure of the Earth

Seismic waves are vibrations in the Earth, which can cause massive destruction.

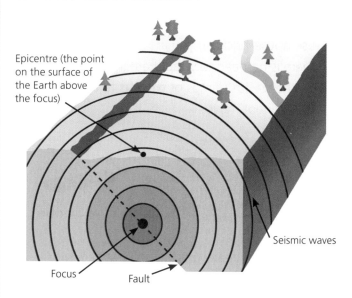

Seismometers detect the vibrations from earthquakes. Simple mechanical devices that consist of a heavy mass, freely suspended, can be used for this. A sheet of paper wrapped around a rotating drum records the vibration.

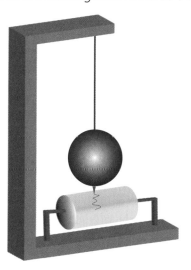

This data is compared to that recorded by other seismometers located in places at known distances from each other. Measurement of the time taken for the waves to arrive allows the origin (focus) of the earthquake to be calculated.

The paths that the waves follow and their speed of travel provide evidence of the Earth's layered structure.

Investigating Earthquakes

You can model the movement of tectonic plates by doing the following:

1. Get two heavy blocks of wood and wrap sandpaper around them. (Alternatively, you can use house-bricks. If you are using house-bricks, there is no need to wrap them in sandpaper.)
2. Fix them with clamps or weights to the bench so they cannot move.

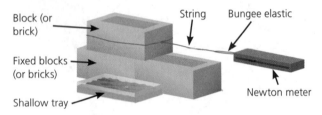

3. Take a third heavy block, also wrapped with sandpaper, and place it on the other two blocks as shown.
4. Measure the third block's position.
5. Tie about 2m of strong string around the block.
6. Attach some bungee elastic to the other end of the string; connect that to a newton meter.
7. Place a small tray of water next to the lower blocks.
8. Pull the top block with the newton meter in a steady manner. Note the meter reading when the block starts to move and the distance it moves.
9. When the block moves, the water surface will be disturbed. This acts like a seismometer. The force multiplied by the distance moved (in metres) gives the energy involved (in joules).
10. Repeat this a few times. The energy does not change in a steady, predictable way. A histogram will show this visually.

Primary and Secondary Waves

When an earthquake occurs, two types of seismic wave are generated.

Primary Waves (P Waves) are detected first. Primary waves are longitudinal waves: the ground is made to vibrate parallel to the direction of the wave. They can travel through solids and liquids and through all layers of the Earth.

Secondary Waves (S Waves) are detected later than P waves, because they travel more slowly. Secondary waves are transverse waves: the ground is made to vibrate at right angles to the direction the wave is travelling. They can travel through solids but not liquids. They cannot travel through the Earth's outer core.

Both P and S waves can be reflected and refracted at boundaries between the crust, mantle and core.

HT When an earthquake occurs, waves radiate outward. The waves change direction gradually (refract) since the density of the rocks increases with depth.

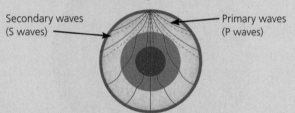

Waves that pass down towards the centre of the Earth meet the boundary between mantle and outer core. S waves are reflected but P waves are refracted. P waves are detected on the opposite side of the Earth to the earthquake but S waves are not.

A study of seismic waves indicates that the Earth is made up of:
- a thin crust
- a mantle which is semi-fluid and extends almost halfway to the centre
- a core which is over half of the Earth's diameter with a liquid outer part and a solid inner part.

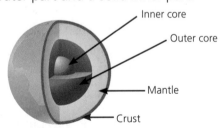

As primary waves are the only type of wave to reach the opposite side of the Earth, and can travel through a liquid, this provides good evidence that the outer core of the Earth is liquid.

P1 Topic 5: Generation and Transmission of Electricity

This topic looks at:
- how to calculate the use and cost of electricity
- how electricity is generated in the classroom and in the power station
- what a transformer does

Electrical Power

An electrical circuit consists of a source of **voltage**. This is either a cell or battery, which transfers energy to charges flowing around the circuit. Voltage is an electrical pressure that gives a measure of the energy transferred. The rate of flow of charge is the **current**.

The electrical energy is transferred to an appliance or device. Power is measured in **watts** (W) and can be calculated from the following formula:

| Electrical power (watt, W) | = | Current (ampere, A) | × | Potential difference (volt, V) | $\dfrac{P}{I \times V}$ |

where *I* is the current

Example

An electric iron draws a current of 4A from the mains supply of 240V. What is its power?

Power = Current × Potential difference

$$= 4 \times 240$$

$$= \mathbf{960W}$$

1 watt is the transfer of 1 **joule** of energy in 1 second.

The **power** of an appliance or device is the amount of energy transferred per second.

| Power (watt, W) | = | $\dfrac{\text{Energy used (joule, J)}}{\text{Time taken (second, s)}}$ | $\dfrac{E}{P \times t}$ |

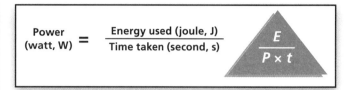

HT ### Example

A computer monitor with a power rating of 200W transfers 200J/s.

How much energy is used by the computer monitor if it is switched on for 30 minutes? (Remember to convert time from minutes to seconds.)

$$P = \frac{E}{t}$$

$$E = P \times t$$

$$= 200 \times 30 \times 60$$

$$= \mathbf{360\,000J}$$

Investigating Power Consumption

The power consumption of electrical devices available in the school laboratory can be easily investigated. Here is one method:

1️⃣ Set up a circuit similar to the one shown below.

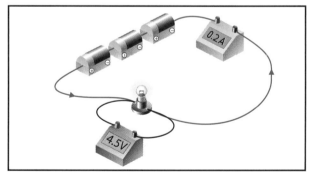

2️⃣ Connect an ammeter in series and connect a voltmeter across the device being investigated (for example, a light bulb, as shown in the diagram).

3️⃣ Record the potential difference and current.

4️⃣ Calculate the power of the device from:

| Electrical power (watt, W) | = | Current (ampere, A) | × | Potential difference (volt, V) |

5️⃣ Repeat for different devices.

Cost of Using Electricity

Energy from the mains supply is measured in kilowatt-hours (kWh), often called a unit. If an electrical appliance transfers 1kWh of energy, it transfers the equivalent of 1 kilowatt (1000W) of power for 1 hour.

To calculate the energy transferred in kWh, the time must be in hours and the power in kW. For example:

- A 200W (0.2kW) television transfers 1kWh of energy if it is switched on for 5 hours.
  0.2kW × 5 hours = 1kWh
- A 2000W (2kW) kettle transfers 1kWh of energy if it is switched on for $\frac{1}{2}$ hour (30 min).
  2kW × $\frac{1}{2}$ hour = 1kWh

To calculate the cost, we need to use the formula:

Cost (p)	=	Power (kilowatts, kW)	×	Time (hour, h)	×	Cost of 1 kilowatt-hour (p/kWh)

Example
A 2000W electric hot plate was used for 90 minutes. How much did it cost to use if 1kWh (unit) costs 14p?

2000W = 2.0kW, 90 minutes = 1.5 hours.

Cost = Power × Time × Cost of 1kWh
$$= 2 \times 1.5 \times 14$$
= 42p

Energy Saving

People have many mains electrical appliances in their homes, from washing machines to kettles, toasters, computers and lights. All of these use electrical energy, so it makes sense to use as many low-energy appliances as possible. The more low-energy appliances that are used, the greater the saving in energy and in cost.

Lighting is a large consumer of electricity. A **disadvantage** of low-energy light bulbs is their original cost, which is much higher than for ordinary bulbs. But this is outweighed by the **advantage** of their saving in running costs and energy.

Example
Suppose one 100W light bulb used for approximately 3 hours a day is replaced by a low-energy 20W bulb. The number of kWh saved in one year is:

$$\frac{(100 - 20)}{1000} kW \times 3 \text{ hours} \times 365 \text{ days}$$

= 87.6kWh

If 1kWh costs 14p, then this means (in one year) a saving of:

$$£0.14 \times 87.6$$

= £12.26

So, as long as the low-energy bulb lasts at least one year and costs less than £12.26, then it would be **cost-efficient** as well as being **energy efficient**.

If the low-energy light bulb cost, say, £6.13, then used for 3 hours a day, it would only take 6 months to make savings equal to the original cost.

In order to calculate its cost-efficiency, we need to calculate the payback time:

$$\textbf{Payback time} = \frac{\textbf{Original cost}}{\textbf{Annual saving}}$$

In this example, payback time = $\frac{£6.13}{£12.26}$

= 0.5 year

However, other disadvantages of low-energy light bulbs are:

- the time taken to come to full brightness (new ones on the market are better)
- not all low-energy light bulbs work with dimmer switches.

Below are further examples of energy efficient appliances and their energy savings compared to older models:

- An energy efficient washing machine could save 30% of the electricity used by an older model.
- An energy efficient kettle could save 20%.
- A modern dishwasher could save up to 40%.

Efficiency of Solar Cells

Solar cells are not very efficient devices. The efficiency of transferring light energy from the Sun into electrical energy from the solar cell used to be generally less than 20%. Scientists in the space industry have increased efficiency to over 40% in research conditions.

Solar cells are currently used in sunny areas or remote places where it is more difficult to get other forms of electrical supply. Increasing efficiency, so they can operate effectively with less sunlight, is the key to making them usable across the world.

Making Electricity by Electromagnetic Induction

If you move a wire (or coil of wire) so that it cuts through a **magnetic field** (magnetic lines of force), then a voltage is induced across the ends of the wire. This will cause electrons to flow along the wire, creating an electric current, if the wire is part of a complete circuit.

Moving the wire down through the magnetic field induces a current in one direction.

Moving the wire up through the magnetic field induces a current in the opposite direction.

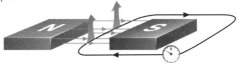

If there is no movement of the wire, there is no induced current.

The same effect can be seen using a coil and a magnet.

1. A coil of wire can be made by wrapping insulated wire around, for example, a pencil. This needs to be connected to a sensitive ammeter, as shown in the diagram below.
2. Move a strong bar magnet towards and away from the coil.
3. Note what happens to the reading on the ammeter.
4. Hold the magnet still. Again, note the reading on the ammeter.
5. Try moving the magnet more quickly and see if there is any difference in the reading.

Moving the magnet into the coil induces a current in one direction.

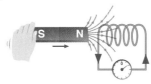

Moving the magnet out of the coil induces a current in the opposite direction.

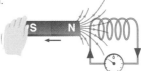

If there is no movement of magnet or coil, there is no induced current.

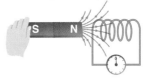

Increasing Voltage and Current

A coil of wire is rotated in a magnetic field. As the coil cuts through the magnetic field, a current is induced in the wire. This current reverses direction every half turn. To increase the voltage, it is necessary to cut through more magnetic field lines per second. This can be done by using stronger magnets, having more coils of wire or moving the wire (or magnet) faster.

Generators use the principle of moving (rotating) coils of wire in a magnetic field to generate electricity; a coil of wire cuts through a magnetic field to induce a voltage. The same effects can be achieved by rotating a magnet within a coil of wire. This method is used in a **bicycle dynamo** (generator) to generate electricity for the bicycle's lights.

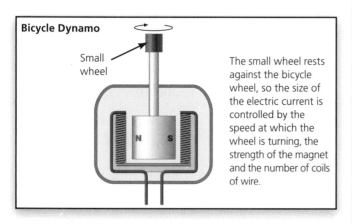

Bicycle Dynamo

Small wheel

The small wheel rests against the bicycle wheel, so the size of the electric current is controlled by the speed at which the wheel is turning, the strength of the magnet and the number of coils of wire.

N S

Types of Electric Current

Direct Current (d.c.)

Direct current flows in one direction only. Cells, batteries and solar cells produce d.c.

In circuit drawings, arrows show direct current flowing from + to –. (However, it is now known that electrons flow from – to +.) The red line on the cathode ray oscilloscope shows d.c.:

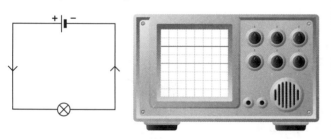

Alternating Current (a.c.)

Alternating current oscillates (reverses its direction) continuously. Mains electricity is a.c. (it has a frequency of 50 hertz). Generators supply alternating current.

50 hertz means the current oscillates 50 times per second. Since it changes direction, you cannot use arrows to show the direction of a.c.. The red line on the cathode ray oscilloscope shows a.c.:

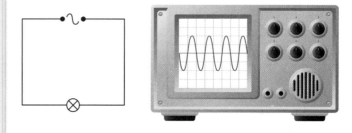

Cells and Batteries

Cells and **batteries** are sources of direct current. A single cell normally gives 1.5 volts. A battery contains two or more single cells (although single cells are commonly referred to as batteries).

Large-scale Electricity Generation

Oil, gas and coal-fired power stations all work in the same way. The boiler burns the fuel and heats water. The high-pressure steam produced is fed into the turbines. This causes the turbine blades to rotate, driving generators. Usually a powerful electromagnet is rotated and a copper coil is kept stationary, much as in the bicycle dynamo. The coil cuts the magnetic field of the magnet and a current is induced.

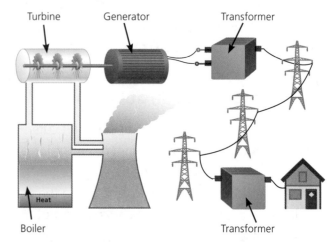

Turbine   Generator   Transformer

Heat

Boiler   Transformer

The Transformer

The electricity from the generator (at about 25 000V) is fed to a **transformer**. Transformers are devices that change the size of an alternating voltage.

Since Power = Current × Voltage, to transmit a large amount of power along the cables from the power station needs either a high current or a high voltage.

A high current means a lot of energy loss in the form of heat because of resistance in the cables, so it is better to increase the voltage and keep the current low.

The transformers **step up** the voltage to 400 000V to transmit the voltage efficiently over the National Grid. Other transformers are then needed to **step down** the voltage at the other end so industry and domestic users can make use of it at safe levels. Since transformers only work on a.c., the electricity from the mains in our homes must also be a.c.

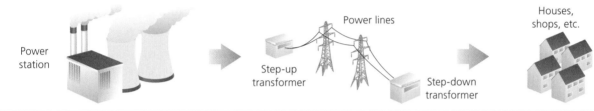

Power station — Step-up transformer — Power lines — Step-down transformer — Houses, shops, etc.

A transformer has a **primary coil** and a **secondary coil**.

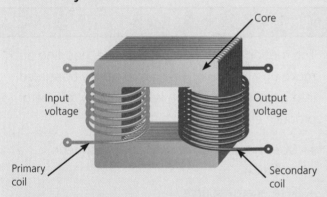

Core — Input voltage — Output voltage — Primary coil — Secondary coil

The ratio of the number of turns of wire on the primary coil N_P to the number of turns on the secondary coil N_S is $\frac{N_P}{N_S}$. This ratio allows us to calculate the size of the output voltage from the secondary coil if we know the primary voltage.

$$\frac{N_P}{N_S} = \frac{\text{Primary voltage}}{\text{Secondary voltage}}$$

If N_S is less than N_P the output voltage will be smaller than the input voltage (step-down transformer).

If N_S is greater than N_P the output voltage will be larger than the input voltage (step-up transformer).

Example 1
The number of turns on the primary coil is 100 and the number of turns on the secondary coil is 1000. The primary voltage is 100V. What is the secondary voltage?

$$\frac{N_P}{N_S} = \frac{100}{1000} = \frac{1}{10}$$

So $\frac{\text{Primary voltage}}{\text{Secondary voltage}} = \frac{1}{10}$

Secondary voltage = 100 × 10

= 1000V

Example 2
A primary coil has 1000 turns. If the primary voltage is 240V, how many turns should the secondary coil have if the output voltage required is 24V?

$$\frac{\text{Primary voltage}}{\text{Secondary voltage}} = \frac{240}{24} = 10$$

So $\frac{N_P}{N_S} = 10$

$\frac{1000}{N_S} = 10$

$N_S = \frac{1000}{10}$

= 100

There are 100 turns on the secondary coil.

Advantages and Disadvantages of Electricity Generation Methods

Fossil Fuels

Coal-fired power stations generate electricity by burning coal to heat water and produce steam, which drives turbines and rotates a generator. Coal is a **non-renewable energy source**.

Oil and gas power stations work in a similar way. There is only a finite amount of these fossil fuels available (which have to be extracted from the Earth and transported). Fossil fuels emit carbon dioxide, which contributes to global warming. There are moves to make the use of coal more acceptable by reducing or recovering the carbon dioxide.

Nuclear Power

Nuclear power stations provide heat to produce steam from nuclear fission. The nuclear material still has to be mined and transported but there is no emission of greenhouse gases. However, nuclear power stations are costly to build and the disposal of radioactive waste poses a problem.

Fossil fuel and nuclear power stations provide most of our electricity (93% in 2010). They are reliable and are quite efficient. As much as 40% of the input energy is converted to electricity.

Renewable Energy Sources

All renewable energy sources (except solar) produce electricity by driving turbines directly. Apart from biomass, they do not produce any atmospheric pollution. However, they often have high initial start-up costs and may have significant impact on their surroundings. Some of the major advantages (+) and disadvantages (−) of using renewable energy sources are given below.

Wind	Waves
The force of the wind turns the blades of a wind turbine, causing a generator to spin and produce electricity.	The motion of the waves makes the 'nodding duck' move up and down. This movement is translated into a rotary movement, which turns a generator to produce electricity.

Wind

+ Does not produce waste or atmospheric pollution.
+ Free energy source.
− Equipment is expensive to install.
− Low output per turbine.
− Wind is unreliable.
− Visual pollution.

Waves

+ Does not produce waste or atmospheric pollution.
+ Free energy source.
− Equipment expensive to install.
− Variable wave size means unreliable, low output.
− Changes appearance of coastline and is a hazard to ships.

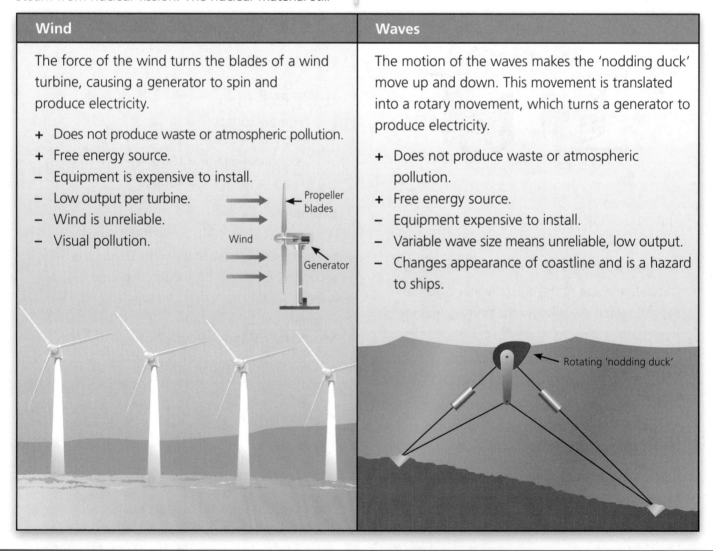

Pumped Storage Hydro-electricity

Water stored in reservoirs above the power station flows down to drive turbines to generate electricity. It is pumped back up when demand is low.

+ Does not produce waste or atmospheric pollution.
+ Reliable and free energy source.
+ Fast response; can support the National Grid during high demand.
+ High output: the water can be used many times.
− Damages habitats and villages.
− Requires high rainfall and mountainous region.
− Changes appearance of surroundings.

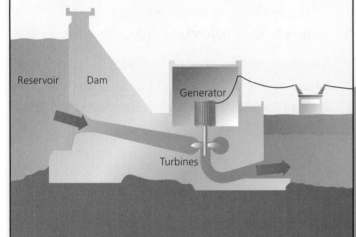

Reservoir   Dam

Generator

Turbines

Solar Power

Solar cells use modern technology to transfer sunlight directly into useful electricity, e.g. in calculators, watches and garden lighting, as well as more sophisticated uses in space probes and satellites.

+ Does not produce waste or atmospheric pollution.
+ Can be used on very small scale, e.g. calculators.
+ No need for turbines and generators.
+ Can be very light, easily portable.
+ Free energy source.
− Can only operate during daylight hours.

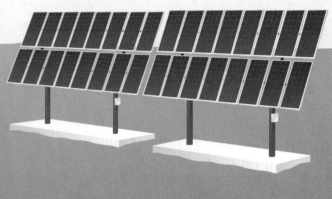

Hazards of Electricity Transmission

It is cheaper to have power lines overhead, rather than putting them underground. However, there is a danger that low-flying aircraft and birds may collide with them during flight.

Substations that house transformers are about 200m apart in urban areas. They often consist of a grey metal box in a fenced enclosure. They all have a yellow 'Danger of Death' sign to warn the public of the real risks of electrocution.

Over the past few years there have been numerous studies to investigate whether there is any risk to the health of people who live near or under electricity power lines. There has been speculation that exposure to low-level magnetic fields might be harmful but, so far, no evidence for this has been found.

Danger of death

P1 Topic 6: Energy and the Future

This topic looks at:
- how to recognise energy transfers in different devices
- the formula to work out efficiency
- how an object radiates and absorbs energy

Energy Transfer

Conservation of Energy

There are many forms of **energy**: thermal (heat) energy, chemical energy, nuclear energy, light energy, sound energy, kinetic energy and potential energy, etc.

The principle of the **conservation of energy** says that energy cannot be made or lost, only transferred from one form into another. For example, the energy of light from the Sun can be transferred to chemical energy by plants through the process of photosynthesis.

Transferring Energy

A roller coaster's energy is constantly transferring between gravitational potential energy (GPE) and kinetic energy (KE).

1. On most roller coasters, the cars start high up with a lot of gravitational potential energy (or they are lifted mechanically, building up gravitational energy).
2. As the cars drop, the gravitational potential energy is gradually being transferred into kinetic energy.
3. The car accelerates to reach its highest speed (maximum kinetic energy) at the bottom of the slope.
4. As the car climbs the slope on the other side, kinetic energy is transferred back into gravitational potential energy.

The height of any other hills or loops in the ride will always be less than the height of the initial one because some kinetic energy is transferred to heat and sound.

Maximum GPE – lowest speed

Maximum GPE

Initial lift to maximise GPE

Cars speed up – GPE transfers into KE

Cars speed up – GPE transfers into KE

Cars slow down – KE transfers into GPE

Maximum KE – greatest speed

A man pushing a car is using **chemical energy** stored in his muscles, which is transferred into **kinetic energy** of the car.

Chemical energy → Kinetic energy

An archer, in drawing back a bow, uses stored **chemical energy**, which is transferred into **elastic potential energy** as the bow is stretched, and then into **kinetic energy** as the bow springs back and releases the arrow.

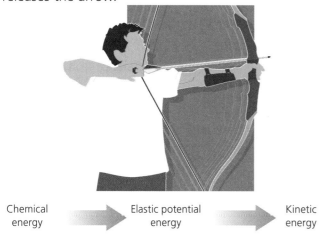

Chemical energy → Elastic potential energy → Kinetic energy

A fossil fuel power station transfers **chemical energy** (stored in the fuel) into **thermal** (heat) **energy** then into **kinetic energy** and finally into **electrical energy**.

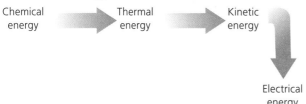

Chemical energy → Thermal energy → Kinetic energy → Electrical energy

The energy transfers in a nuclear power station are nearly the same except it is **nuclear energy** that is transferred into **thermal energy** to begin with.

A wind turbine transfers **kinetic energy** into **electrical energy**.

Kinetic energy → Electrical energy

Household appliances such as an electric kettle or iron transfer electrical energy into thermal energy.

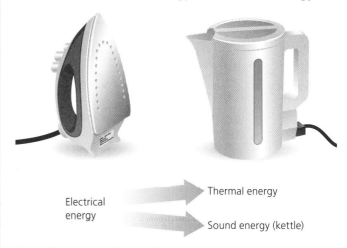

Electrical energy → Thermal energy
Electrical energy → Sound energy (kettle)

Note that since the kettle makes a noise as it heats the water, sound energy is also transferred.

Light bulbs (even low-energy light bulbs) become warm, or even hot, so they transfer electrical energy into light and thermal energy.

Electrical energy → Light energy
Electrical energy → Thermal energy

N.B. **Energy transfer diagrams** shown above illustrate a convenient way to represent the energy transfer that is involved.

Efficiency

When devices transfer energy, only part of it is usefully transferred to where it is wanted and in the form that is wanted. The remainder is 'wasted'.

The proportion of useful energy transferred by an appliance is called the **efficiency** of the appliance and is calculated using:

$$\text{Efficiency} = \frac{\text{Useful energy transferred by the device}}{\text{Total energy supplied to the device}} \times 100\%$$

N.B. No device can have an efficiency greater than 100%.

Example – Electric Kettle

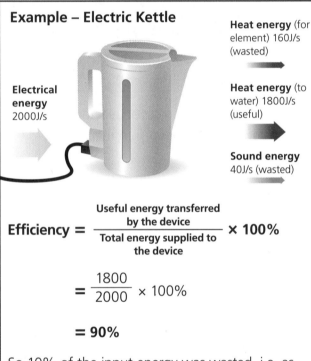

Heat energy (for element) 160J/s (wasted)

Electrical energy 2000J/s

Heat energy (to water) 1800J/s (useful)

Sound energy 40J/s (wasted)

$$\text{Efficiency} = \frac{\text{Useful energy transferred by the device}}{\text{Total energy supplied to the device}} \times 100\%$$

$$= \frac{1800}{2000} \times 100\%$$

$$= 90\%$$

So 10% of the input energy was wasted, i.e. as heat and sound. So if you add up the total energy output in 1 second, it is:

$$\underset{\substack{\text{(wasted} \\ \text{heat)}}}{160} + \underset{\substack{\text{(useful} \\ \text{heat)}}}{1800} + \underset{\substack{\text{(wasted} \\ \text{sound)}}}{40} = 2000J$$

$$= \text{input}$$

The total amount of energy before the transfer is therefore equal to the total amount of energy after the transfer. This agrees with the **principle of conservation of energy** and is true for any appliance that transfers energy.

Radiation and Absorption of Energy

A hot object radiates thermal (heat) energy. It transfers energy to the environment, so it cools down. If the environment is hotter than the object, the object absorbs thermal energy and heats up. For the object to stay at constant temperature, the rate of energy **absorption** (average power) equals the rate of energy radiated. The amount of energy radiated or absorbed depends on the colour and texture of the surface.

Investigating the Effect of a Surface on Energy

Hot water

Thermometer

Shiny white

Matt black

A Leslie Cube is a hollow metal cube with white shiny and matt faces, and black shiny and matt faces. To investigate **radiated energy**, the cube is filled with hot water. A thermometer is set up to record the temperature after a fixed time and at a fixed distance from each face of the cube. Comparing the results shows the effect of the surface on the amount of energy radiated. To investigate how the surface affects **absorbed energy**, the cube can be filled with cold water.

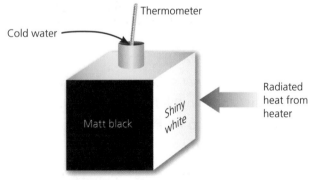

Thermometer

Cold water

Matt black

Shiny white

Radiated heat from heater

A radiant heater (electrical or Bunsen burner) is placed near to one face and the temperature measured with a thermometer. This is repeated with fresh cold water in the cube but with a different face placed at the same distance from the heater. A **matt black** surface will both radiate and absorb thermal energy faster than a surface that is light coloured and shiny.

Questions labelled with an asterisk (*) are ones where the quality of your written communication will be assessed – you should take particular care with your spelling, punctuation and grammar, as well as the clarity of expression, on these questions.

1 Give one similarity and two differences between longitudinal and transverse waves. **(3)**

2 Describe how to measure the focal length of a converging lens. **(4)**

3 What are the advantages of using a reflecting telescope compared to a simple telescope? **(3)**

4 Explain the following terms used in astronomy:

(a) the Solar System **(1)**

(b) a galaxy **(1)**

(c) the Milky Way **(2)**

5 (a) What is meant by a 'light year'? **(1)**

(b) Why do astronomers use the measurement 'light year' when they refer to stars? **(2)**

6 What are the two ways to look for intelligent life in the Universe? **(2)**

7 Give two pieces of evidence to support the Big Bang Theory. **(2)**

8 Name two uses of:

(a) ultrasound **(2)**

(b) infrasound. **(2)**

9 A fishing trawler is searching for a shoal of fish. It sends an ultrasound wave into the water in the direction of the fish. This takes 2s to be transmitted, reflected and received back at the trawler. How deep should the trawler expect to find the fish?
(Use the equation Distance (m) = Wave speed (m/s) × Time (s), $x = v \times t$)
Take the speed of sound in water to be 1500m/s. **(2)**

10 *Describe how a tsunami may be caused by an earthquake. **(6)**

11 An electric heater is rated at 1000W.

(a) How much energy does it use in 1 hour?
(Use the equation Energy used (J) = Power (W) × Time taken (s), $E = P \times t$) **(1)**

(b) How much does it cost to use the heater for 90 minutes if a unit of electricity costs 16p? **(1)**

12 A low-energy light bulb saves energy and money. Give one disadvantage of a low-energy light bulb. **(1)**

13 Why is a transformer used at a power station? **(2)**

14 A hairdryer wastes some of the electrical energy supplied.

(a) Explain what this means. **(2)**

(b) 1200J/s of electrical energy is supplied. 120J/s of this is wasted. Calculate the hairdryer's efficiency. **(2)**

15 What transfers of energy occur when:

(a) you speak to someone on the phone? **(2)**

(b) you throw a ball straight up in the air? **(2)**

16 Tony puts two similar shirts out to dry in the sun after washing them. One is a dark colour, the other is white. After an hour he goes to see if they are dry. One of them is but the other is still damp. Which one would you expect to be dry? Explain your answer. **(3)**

17 What is the difference between the geocentric and heliocentric models of the Solar System? **(1)**

18 What is 'red-shift'? **(1)**

HT **19** Explain what happens to waves refracted at a boundary from one material to another in terms of their speed and direction. **(3)**

20 X-rays have a lower frequency than gamma radiation so they are safer to use. Discuss the accuracy of this statement. **(3)**

21 (a) Our Sun will at some stage become a red giant. Explain how it will eventually become a black dwarf. **(4)**

*(b)** Describe in detail the evolution of a star with a mass much larger than the mass of our Sun. Explain any terms that you use. **(6)**

22 What is meant by 'red-shift'? Why does it provide evidence for the expansion of the Universe? **(4)**

23 When an earthquake occurs, seismic waves are produced.

(a) Why do these waves change direction as they pass through the Earth? **(1)**

(b) Some of the waves that are directed straight down towards the centre of the Earth are reflected. Why is this? **(3)**

(c) (i) What happens when the other type of wave is directed downwards, to the centre of the Earth? **(1)**

(ii) What is the evidence for this? **(1)**

(iii) What can be deduced from this about the structure of the inside of the Earth? **(2)**

24 Why do scientists find it difficult to predict when an earthquake will happen? **(2)**

25 A student wants to construct a transformer to step down a voltage of 24V to 4V. She winds 300 turns onto a primary coil. How many turns of wire should she wind on to the secondary coil? **(2)**

26 An electrical device is 86% efficient. Its power is rated at 1.2kW. How much useful energy is transferred every second? **(2)**

Answers

Model answers have been provided for the quality of written communication questions that are marked with an asterisk (*). The model answers would score the full 6 marks available. If you have made most of the points given in the model answer and communicated your ideas clearly, in a logical sequence with few errors in spelling, punctuation and grammar, you would get 6 marks. You will lose marks if some of the points are missing, if the answer lacks clarity and if there are serious errors in spelling, punctuation and grammar.

Unit B1

1.*(a) Fish have wet scales and gills. They lay eggs in water and are cold blooded. Amphibians have smooth, moist, permeable skin. Adults have lungs whilst young have gills. They lay eggs in water or damp places and are cold blooded. Reptiles have dry, scaly skin. They lay eggs and are cold blooded. Birds have feathers and a beak. They lay eggs and are warm blooded. Mammals have hair and produce milk. They give birth to live offspring. They have lungs and are warm blooded.

(b) Vertebrates have a backbone; invertebrates do not

2. (a) **Any suitable answers (2 marks for characteristics, 2 marks for reasons), e.g.** Camels have a large surface area to volume ratio to increase heat loss **(1 mark)** and a sandy coat so they are camouflaged **(1 mark)**; Fish have a streamlined shape to travel quickly through water **(1 mark)** and gills with a large surface area to absorb oxygen from water **(1 mark)**

(b) Habitats with conditions outside those where most organisms live **(1 mark)**. **Any suitable example, for 1 mark, e.g.** Hydrothermal vents

(c) (i) Variation **(1 mark)**

(ii) Arises through genetic factors **(1 mark)** and environmental factors **(1 mark)**

(d) Continuous variation: height or weight **(1 mark)**; Discontinuous variation: blood group or eye colour **(1 mark)**

3.*Individuals in a population show variation. Most organisms produce more young than will survive to adulthood. Population sizes are generally stable, so there is competition between organisms. Individuals with advantageous characteristics are more likely to survive. These organisms are more likely to reproduce and pass on their characteristics. Over time more individuals in a population will possess the advantageous characteristics.

4. (a) D

(b)(i) **Any suitable answer, e.g.** Cystic fibrosis; Sickle cell disease

(ii) **Any two symptoms for the disorder given, e.g.** Cystic fibrosis: Being unable to digest food **(1 mark)**; Airways clogged with mucus **(1 mark) or** Sickle cell disease: Shortness of breath **(1 mark)**; Dizziness **(1 mark)**

(c)

	E	E
e	Ee	Ee
e	Ee	Ee

5. (a) Maintaining a stable internal environment

(b) **Any suitable answer, e.g.** Blood vessels dilate causing greater heat loss **(1 mark)**; Sweat glands release sweat which evaporates and cools us **(1 mark)**; Erector muscles cause hair to lie flat and not trap heat **(1 mark)**; Sebaceous glands produce oily sebum to encourage sweat to spread effectively **(1 mark)**

6. (a) C

(b) Sensory neurones: take nerve impulses from the sense organs to the central nervous system **(1 mark)**; Relay neurones: pass nerve impulses from the sensory neurones to the motor neurones **(1 mark)**; Motor neurones: take nerve impulses from the central nervous system to the muscles or glands **(1 mark)**

Answers

(c) Nerve impulse reaches a synapse and releases a neurotransmitter **(1 mark)**; Neurotransmitter crosses the synapse **(1 mark)**; which starts a nerve impulse in the next neurone **(1 mark)**; Neurotransmitter then destroyed or removed **(1 mark)**

7. (a) Pancreas produces insulin **(1 mark)**; which converts glucose from the blood to glycogen **(1 mark)**; Glycogen is removed from the blood and stored in the liver **(1 mark)**

(b) **Any four from:** Increased urination; Increased thirst; Increased tiredness; Weight loss; Blurred vision

(c) Type 1 diabetes: cannot produce insulin; Type 2 diabetes: resistant to the insulin produced

(d) Eat three meals a day; Include carbohydrate in the diet but reduce fat and sugar; Be physically active

8. B

9. (a) Substances that affect the central nervous system, causing changes in psychological behaviour and possibly addiction

(b) **1 mark for effect, 1 mark for example, e.g.** Stimulants: speed up the transmission of messages across synapses, e.g. Caffeine; Depressants: slow down the transmission of messages across synapses, e.g. Alcohol

10. (a) **Any three (correctly identified as organ or tissue) for 1 mark each:** Organs: heart, kidney, liver, lung, pancreas; Tissues: bones, tendon, cornea, skin, heart valves, veins

(b) **Any three suitable answers, e.g.** Some people believe that we should be forced to give up our organs upon death (and not volunteer); Others believe that people who may have had some control over their condition should not be allowed transplants (e.g. liver transplants for alcoholics); Organs are sometimes stolen and trafficked

11. (a) Bacteria; Fungi; Viruses

(b) Organisms that transmit pathogens **(1 mark)** **Any suitable example for 1 mark, e.g.** the mosquito for malaria

(c) Physical barriers **(any two for 1 mark each)**: skin; mucus; cilia. Chemical barriers: lysozyme enzymes in tears **(1 mark)**; stomach acid **(1 mark)**

12. (a) The dynamic relationship between living things

(b) **2 marks for any suitable food chain, e.g.** Grass → rabbit → stoat → fox. **2 marks for appropriate key words, e.g.** Producer; Herbivore; Carnivore; Top carnivore

(c) Higher levels of a pyramid of biomass are always smaller than those below them

(d) **2 marks for descriptions, 2 marks for examples, e.g.** Parasitic relationships: benefit infecting organism at the expense of host, e.g. Fleas infecting humans; Mutualistic relationships: benefit both infecting organism and host, e.g. Oxpecker birds that pick ticks from the hides of large mammals

13. **Any two substances from air (1 mark each) and two from water (1 mark each) with sources identified correctly, e.g.** Air: Hydrocarbons; Carbon dioxide; Sulfur dioxide – all from burning fossil fuels; Carbon monoxide from vehicle exhausts. Water: Sewage from human waste; Nitrates from fertilisers; Phosphates from water from laundries and from fertilisers

14.*Photosynthesis in plants removes carbon dioxide from the atmosphere and stores it as glucose. When plants respire, carbon dioxide is released back into the atmosphere. When plants are eaten, carbon is passed along the food chain. When animals respire, carbon dioxide is released into the atmosphere. When plants and animals die, microorganisms feed on their remains, releasing carbon dioxide. In addition, burning fossil fuels releases carbon dioxide into the atmosphere.

Answers

15. It allows organisms to be identified **(1 mark)**; which makes it easier to study habitats or species that need conservation **(1 mark)**

16. 'Faulty' recessive alleles are inherited from both parents

17. **(a)** A reduction in blood flow to the surface of the skin keeps the body warmer
    (b) An increase in blood flow to the surface of the skin allows more heat to be lost

18. Pancreas produces glucagon **(1 mark)**; which converts glycogen from the liver into glucose **(1 mark)**; Glucose is released into the blood from the liver **(1 mark)**

19. **Any three from:** Selective weedkillers disrupt the growth of all plants except grass **(1 mark)**; Root powder promotes root growth in cuttings **(1 mark)**; Seedless fruit can be grown all year around **(1 mark)**; Ethene is used to ripen fruit **(1 mark)**

20. **(a)** MRSA have evolved to become immune to some antibiotics
    (b) As a result of misuse or overuse of antibiotics

21. *Nitrogen-fixing bacteria convert atmospheric nitrogen into nitrates in the soil. These are taken up by plants to make protein. When plants are eaten the nitrogen in the plant protein is passed along the food chain. Decomposers break down dead animals and plants. Soil bacteria convert proteins and urea into ammonia. Nitrifying bacteria convert ammonia into nitrates in the soil. Denitrifying bacteria convert nitrates into atmospheric nitrogen. Lightning can also convert nitrogen gas into nitrates.

Unit C1

1. **(a)** B
   (b) A salt

2. **(a)** The Earth's early atmosphere was mostly composed of carbon dioxide **(1 mark)**; Over time levels of carbon dioxide decreased and levels of nitrogen and oxygen increased **(1 mark)**
   (b) When the Earth and its atmosphere cooled, water vapour released into the atmosphere by volcanic eruptions condensed to become liquid **(1 mark)**; This fell as rain and filled up the hollows in the Earth's crust **(1 mark)**

3. **Any three from:** Through burning fossil fuels; Deforestation; An increase in cattle farming / rice growing; Motorised transportation

4. **(a)** A compound made up of hydrogen and carbon only
   (b) **Any one from:** A greater molecular mass; Higher melting/boiling point; More viscous; Less easy to ignite
   (c) Because it contains lots of different hydrocarbon molecules of different sizes
   (d) B

5. **(a)** **Any one from:** Cuts down on excavation / mining; Uses less water and chemicals; Saves on raw materials; Less energy required
   * **(b)** Recycling more materials such as metals, paper, glass and plastics can be considered to be sustainable because it helps to conserve natural resources, it reduces the amount of energy needed to turn the used material into a new product, and it provides more employment opportunities within the ever-increasing recycling industry. By recycling these materials there will be less future need in economic investment to find new replacement materials or new sources of the material. This also means that the metals, paper, glass and plastics will not be left on a landfill site, causing potential environmental risks and damaging any future use of that area.

Answers

6. (a) An ore is a naturally occurring mineral in the Earth's crust that contains compounds of metals.

(b) A metal's position in the reactivity series determines how easily it is extracted.

(c) (i) Copper; Carbon dioxide

(ii) It has been reduced to copper metal.

(d) Loss of oxygen or gain of electrons in a compound during a chemical reaction.

7. Aluminium can only be extracted by electrolysis **(1 mark)**; Aluminium oxide cannot be reduced by carbon because carbon is less reactive than aluminium **(1 mark)**; Iron is extracted by heating the iron oxide with carbon **(1 mark)** because carbon is more reactive than iron **(1 mark)**

8. (a) Calcium oxide; Carbon dioxide

(b) If heated with clay, it will produce cement, which can be used to make concrete or plaster **(1 mark)**; If heated with sand, it will produce glass **(1 mark)**

(c) At least one point from environmental, economic and social, plus any additional point, from: Environmental: The effect on native animal habitats / landscape; Noise and air pollution; Additional trafficEconomic: Costs involved in quarrying and processing; Effect on local businesses Social: Availability of workforce; How the quarry can be used afterwards

9. (a) Sulfuric acid + Magnesium oxide ⟶ Magnesium sulfate + Water

(b) Nitric acid + Copper oxide ⟶ Copper nitrate + Water

(c) The stomach contains hydrochloric acid, which reacts with the base **(1 mark)** in the indigestion remedy to form a salt and water **(1 mark)**

(d) To kill harmful bacteria on any food consumed **(1 mark)**; To give enzymes their optimum environment for digestion **(1 mark)**

10.(a) A monomer is a short-chain hydrocarbon **(1 mark)**; A polymer is a large molecule made up of a repeating pattern of identical smaller chemical molecules called monomers **(1 mark)**

(b)
$$\left[\begin{matrix} F & F \\ | & | \\ C & - C \\ | & | \\ F & F \end{matrix}\right]_n$$

(c) B

(d) Burning polymers produces air pollution **(1 mark)**; The carbon dioxide that is also produced contributes to the greenhouse effect **(1 mark)**; When burned some polymers produce toxic fumes **(1 mark)**

(e) Any one from: Water-resistant; Flexible; Non-stick; Unreactive

11. (a) Ions in the brine are free to move around whilst in solution **(1 mark)**; Electric (d.c.) current passed through the electrolyte causes positive hydrogen ions to flow to the negative electrode forming hydrogen gas **(1 mark)**; and negative chloride ions to flow to the positive electrode forming chlorine gas **(1 mark)**

(b) (i) A greenish-yellow gas

(ii) Test using damp litmus paper, which will be bleached by the chlorine

(c) Gas is poisonous **(1 mark)**; Would cause an environmental hazard if large quantities released into the atmosphere **(1 mark)**

12.* The atoms in pure iron are all the same size and in a regular arrangement. By mixing iron with other elements it becomes an alloy. In an alloy the regular arrangement of the iron atoms (ions) is disrupted because the atoms of the alloying element atoms are either larger or smaller than the iron atoms (ions). This gives alloys improved properties over the pure metal, for example, mixing carbon with iron makes steel. In steel the carbon atoms interrupt the layers of iron atoms (ions) so they are no longer able to move easily over each other, making the alloy much harder and stronger so it can be used in the construction industry.

Answers

13. (a) (i) $CH_4(g) + 2O_2(g) \longrightarrow CO_2(g) + 2H_2O(l)$

(ii) For example: Methane + Oxygen $\longrightarrow$ Carbon dioxide + Carbon monoxide + Water **(1 mark)**

$3CH_4 + 5O_2 \longrightarrow CO_2 + 2CO + 6H_2O$ **(1 mark)** (see additional reactions on page 46)

* (b) Supply refers to the amount of petrol or diesel oil that can be obtained from the fractional distillation of crude oil and demand is the amount that is needed by consumers. The demand for petrol is far greater than the amount that can be supplied because it is a shorter chain hydrocarbon that can release energy more quickly by burning so making it a good fuel. In comparison the amount of diesel oil that can be supplied is far greater than the consumers' need for the fuel because it is a longer chain hydrocarbon and does not make as good a fuel as petrol. The shortfall in petrol supply can be overcome by taking the diesel oil that is in excess and cracking the diesel oil into shorter chain hydrocarbons such as petrol.

Unit P1

1. Similarity: They carry energy
Differences: **Any two from:** They travel at different speeds; The patterns of disturbance are different; Longitudinal waves need a medium but some transverse waves do not.

2. Hold lens up to distant light source, e.g. window **(1 mark)**; Catch image on screen (behind lens) **(1 mark)**; Move lens until sharp image seen **(1 mark)**; Measure distance from lens to image/screen **(1 mark)**

3. Mirrors cheaper to make **(1 mark)**; Large mirrors easier to make than large lenses **(1 mark)**; Reflecting telescopes gather much more light **(1 mark)**

4. (a) The Sun and the planets which orbit it
(b) (Collection of) billions of stars
(c) The name of a galaxy **(1 mark)**; The galaxy in which we live/the Solar System is **(1 mark)**

5. (a) The distance light travels in one year
(b) Distance to stars is very large/huge **(1 mark)**; It is easier to write (huge) distances in light years than kilometres **(1 mark)**

6. Sending spaceships; Searching for radio signals

7. Red-shift/galaxies moving away; Cosmic microwave background radiation

8. (a) Any two from: Foetal scans; Imaging, e.g. heart; Sonar; Locate prey/communicate (some animals)
(b) Any two from: Detect meteors; Volcanic eruptions; Locate (some) animals; Use it to communicate (some animals)

9. Depth = 1500 × 1 **(1 mark)** = 1500m **(1 mark)**

10.* The surface of the Earth is split into several large tectonic plates. Convection currents in the mantle cause the plates to move. It is at the boundaries of these moving plates that volcanic, earthquake and mountain-forming zones occur. This is because when the plates slide past each other the movement is not smooth and the plates may get stuck. Over time pressure builds up. Earthquakes are triggered when the tectonic plates that make up the Earth's surface suddenly move, releasing the pressure. A tsunami is usually caused by a powerful earthquake under the ocean floor. This earthquake pushes a large volume of water to the surface, creating waves. These waves are the tsunami. In the deep ocean these waves are small. As they approach the coast the height of the wave increases. A tsunami can also be triggered by a volcanic eruption, landslide or other movements of the Earth's surface.

Answers

11. **(a)** $1000 \times 60 \times 60 = 3\,600\,000$ J
    (b) 24p

12. **Any one from:** High initial cost; Takes time to come to full brightness; Some do not work with dimmer switch

13. Increase / step up voltage; Reduces energy loss

14. **(a)** Produces sound (energy) / makes a noise **(1 mark)**; This is not useful **(1 mark)**
    (b) $\frac{1080}{1200}$ **(1 mark)** $\times 100 = 90\%$ **(1 mark)**

15. **(a)** Sound to electrical **(1 mark)**; Electrical to sound **(1 mark)**
    (b) Chemical to kinetic **(1 mark)**; Kinetic to potential **(1 mark)**

16. Dark coloured **(1 mark)** as the dark colour absorbs heat / thermal radiation **(1 mark)** better / faster than white or light colours **(1 mark)**

17. Geocentric: the Earth at the centre; Heliocentric: the Sun is at the centre

18. Light from distant sources is moved or shifted towards the red part of the visible spectrum.

19. Waves travelling from more dense to less dense medium **(1 mark)** speed up **(1 mark)** and refract away from normal **(1 mark)** (Or less dense to more dense, slow down, refract towards normal)

20. Generally higher frequency means more danger **(1 mark)**; X-rays are still dangerous **(1 mark)**; High-energy X-rays are as dangerous as low-frequency gamma **(1 mark)**

21. **(a)** Red giant will cool **(1 mark)**, then collapse under its own gravity **(1 mark)** to become a white dwarf **(1 mark)**, then a black dwarf **(1 mark)**
    *__(b)__ After the main stable period the star will expand to become a red supergiant. Red supergiants are hundreds of times bigger than the Sun. When the fusion reactions in the star stop, the star collapses. As it collapses it heats up again. This process is repeated until the star becomes unstable. When a supergiant dies, it shrinks rapidly and then it explodes, releasing massive amounts of energy, dust and gas. This is called a supernova. The star's core is still left behind. This core is massive and has two possible fates. It could become a neutron star or a black hole. To become a neutron star it has to be lighter than three solar masses and anything above this may become a black hole. This happens because the core has such a large mass and gravitational force it collapses in on itself. The remaining gas and dust may form new stars.

22. Shift of light frequency to red **(1 mark)** from source moving away **(1 mark)**; Light from galaxies red-shifted **(1 mark)**; Galaxies must be moving away from us **(1 mark)**

23. **(a)** Density of rocks increases with depth
    (b) They are S / secondary / transverse waves **(1 mark)**; They meet the boundary between solid / mantle and liquid / core **(1 mark)**; They cannot travel through liquid / core **(1 mark)**
    (c) (i) Other / P / primary / longitudinal waves refract at a boundary
    (ii) Detected on the opposite side of the Earth
    (iii) Must have travelled through the centre **(1 mark)** so centre / core must be liquid **(1 mark)**

24. Earthquakes occur when plates slide past each other **(1 mark)**; Plates do not move in regular patterns **(1 mark)**

25. $\frac{24}{4} = 6$ **(1 mark)**; $\frac{300}{6} = 50$ turns **(1 mark)**

26. 1200×0.86 **(1 mark)**; $= 1032 = 1030$ J (3sf) **(1 mark)**

Acid rain – rain that has reacted with gaseous pollutants such as sulfur dioxide and nitrogen dioxide. The gases are as a result of fossil fuels being burned.

Alkane – a saturated hydrocarbon that contains only single carbon–carbon covalent bonds. The number of hydrogen atoms is double the number of carbon atoms plus two.

Alkene – an unsaturated hydrocarbon that contains one double covalent carbon–carbon bond. The number of hydrogen atoms is double the number of carbon atoms.

Alleles – alternative forms of the same gene.

Alloy – a mixture of two or more metals, or a mixture of one metal and a non-metal.

Alternating current (a.c.) – current that continuously reverses its direction.

Amplitude – the maximum vertical disturbance caused by a wave.

Atom – smallest particle of a chemical element that can exist.

Big Bang – the rapid expansion of material at an extremely high density; the event believed by many scientists to have been the start of the Universe.

Biodiversity – the variety of different types of organisms in a habitat or ecosystem.

Biofuel – a source of renewable energy that is made from biological materials that include plants and animal waste.

Biomass – the total mass of organic material (no water) of an organism or population.

Calcium carbonate – compound that has the formula $CaCO_3$ and is found in many different types of rock.

Calcium hydrogencarbonate – compound that has the formula $Ca(HCO_3)_2$ and is found dissolved in water. Also known as calcium bicarbonate.

Carbon cycle – the process of carbon being recycled between living organisms and the environment.

Cement – a building material made by grinding calcined limestone and clay to a fine powder, then heating in a rotary kiln.

Chlorine – halogen with symbol Cl; a yellow, poisonous gas.

Chromosome – made up of DNA and protein; consists of a series of genes.

Classification – groups into which all living things are placed according to their characteristics.

Compound – a substance that contains two or more different atoms of one or more elements chemically combined.

Concrete – used in the construction industry and made by combining cement with sand, aggregate and water.

Conservation of energy – a law that states that energy cannot be made or lost: it can only be transferred from one form into another.

Convection current – the current or cycle of hot gas or liquid rising and cold gas or liquid falling.

Converging lens – a lens thicker in the middle than at its edges; converges light to a real focus.

Cosmic microwave background (CMB) – radiation left over from the Big Bang.

Cracking – the process of breaking down larger hydrocarbons into smaller, more useful ones using a catalyst and heat.

Crude oil – a naturally occurring flammable liquid that is made up of a mixture of hydrocarbons.

Current – the rate of flow of electrons through a conductor (measured in amperes / amps (A), milliamps (mA)).

Decolourise – to lose colour, for example, bromine water will lose its colour when added to an alkene.

Deforestation – the clearance of naturally occurring forests by logging and burning.

Glossary

Diabetes – a medical condition; the pancreas fails to produce enough insulin to keep blood sugar at the correct level.

Direct current (d.c.) – the flow of current in one direction only.

Discharged – in electrolysis this is when the positive or negative ion either gains or loses electron at the electrode to become a neutral atom.

DNA – the substance from which chromosomes are made.

Dominant (allele) – an allele that is expressed even if paired with a recessive allele.

Dynamo – a device for generating electricity from the simple rotational motion of a coil in a magnetic field or the rotation of a magnet inside a coil.

Efficiency – the ratio of the useful energy obtained from a device compared to the amount of energy put into the device to operate it.

Electrical energy – energy of electric charges or current; the product of the voltage (volts), current (amps) and time (s); measured in joules (J).

Electrolysis – splitting up of a molten or aqueous solution of salt, using electricity.

Electromagnetic spectrum – a continuous arrangement that displays electromagnetic waves in order of increasing frequency or wavelength.

Electromagnetic waves – energy waves that make up the electromagnetic spectrum; they are transverse and travel through a vacuum at the speed of light.

Element – a pure substance that is made from just one type of atom. It cannot be chemically split into anything simpler.

Elongation – the lengthening of plant cells caused by the hormones auxin and gibberellin.

Energy – the ability to do work; it can be transferred from one place to another (e.g. along a wire, as electrical energy) and transferred into other types (e.g. from electrical to light).

Ethanol – a clear, colourless alcohol found in beverages such as wine, beer and brandy. It has the formula C_2H_5OH.

Eutrophication – the excessive growth and decay of aquatic plants due to increased levels of nutrients in the water, which results in oxygen levels dropping so fish and other populations die.

Evolution – the changes that take place in living things over a very long period of time as they become better adapted to their environment.

Extreme habitat – a place where the living conditions for organisms are particularly harsh, e.g. the Antarctic (extreme cold).

Extrusive igneous rock – fast cooling of lava at the Earth's surface forms rock with small crystals.

Focal length – the distance from the centre of a lens to its focus.

Fossil fuel – natural source of energy, such as oil, coal or natural gas, that has been made from the remains of plants and animals over millions of years.

Fractional distillation – a method of separating a mixture of liquids that have different boiling points.

Fractions – a group of hydrocarbons with similar boiling points.

Frequency – the number of complete wave oscillations per second, or the number of complete waves to pass a point in 1 second; measured in hertz (Hz).

Gamma rays – high-frequency electromagnetic waves with a short wavelength.

Gene – the part of a chromosome that controls the development of a particular characteristic.

Genotype – the genetic identity of an organism; especially used in monohybrid inheritance.

Glass – a non-crystalline solid material made from heating limestone and silicon dioxide.

Global warming – an increase in the Earth's average atmospheric temperature that causes corresponding changes in climate and that may result from the greenhouse effect.

Heterozygous – different alleles in a genetic pair.

Homeostasis – maintaining a stable internal environment (in our bodies), e.g. maintaining the correct temperature and levels of water.

Homozygous – the same alleles in a genetic pair.

Hormone – a chemical message produced by an endocrine gland.

Hydrocarbons – compounds that are made of only hydrogen atoms and carbon atoms.

Igneous rock – rock that is formed by the cooling and solidification of molten magma.

Indicator species – organisms that can indicate the presence or lack of something, e.g. pollution.

Infrared – a region of the electromagnetic spectrum with wavelengths just beyond the red end of the visible spectrum.

Infrasound – sound with frequency less than 20Hz, below the limit of human hearing.

Inheritance – the passing on of features through reproduction.

Insulin – the hormone produced by the pancreas that lowers blood sugar levels in the body.

Interdependence – the dynamic relationship between all living things.

Intrusive igneous rock – rocks with large crystals formed by the slow cooling of molten magma within the Earth's crust.

Ion – a positively or negatively charged particle formed when an atom or group of atoms lose or gain electrons.

Kinetic energy (KE) – the energy possessed by a moving object; measured in joules (J).

Light year – the distance that light travels in a year.

Longitudinal wave – an energy-carrying wave in which the movement of the particles is in line with the direction in which the energy is being transferred.

Magnetic field – a field of force that exists around a magnetic body.

Metal oxides – formed when a metal ion combines with an oxide ion.

Metamorphic rock – sedimentary or igneous rock that has been changed by the action of intense heat and pressure.

Microwave – a region of the electromagnetic spectrum between infrared and radio waves.

Natural selection – the survival of individual organisms that are best suited / adapted to their environment (the basis of Darwin's theory of evolution).

Nebula – an immense, sometimes luminous cloud of gas and dust in interstellar space, found outside our Solar System; may result from the explosion of a star.

Neurone – a cell that carries nerve impulses.

Neurotransmitters – chemical messages transmitted between ends of neurones at synapses.

Neutralisation – when an alkali is reacted with an acid to form a salt and water. The resultant solution will have a pH of 7.

Nucleus – the part of a cell that contains the chromosomes.

Ore – a type of rock that contains minerals with important elements including metals in sufficient quantities to make it commercially viable to extract.

Oviparous – an animal that lays eggs.

Oxidation – when electrons are lost from a reacting element or compound.

Pathogen – a microorganism that causes a disease.

Phenotype – the physical expression of the genes that an individual possesses, e.g. brown eyes.

Photosynthesis – the process by which green plants use light energy to make glucose.

Glossary

Phototropic (positively) – a plant growth response to light (shoots and stems grow towards the light).

Plaster – a mixture of lime, sand and water, sometimes with fibre added, that hardens to a smooth solid and is used for coating walls and ceilings.

Pure – a substance that is free from foreign elements.

Poikilotherm – a plant or animal species that cannot maintain thermal homeostasis (its body temperature varies with the environment).

Poly(chloroethene) – made from chlorinated ethane molecule and commonly referred to as PVC. It can be used to make windows and doors.

Poly(ethene) – made from the monomer ethene and used to make plastic bags.

Polymer – a large molecule composed of repeating structural units, typically connected by covalent chemical bonds.

Poly(propene) – made from the monomer propene and used to make plastic bottles.

Potential difference (p.d.) – same as voltage: difference in electrical voltage between two points in a circuit; expressed in volts (V).

Potential energy (PE) – the energy stored in an object as a consequence of its position, shape or state (includes gravitational, elastic and chemical); measured in joules (J).

Power – the rate at which work is done or energy is transferred by a device; measured in watts, 1W = 1J/s; also refers to the strength of a lens, measured in dioptres, D.

Primary waves – longitudinal waves generated by an earthquake.

Radiation – the process of transferring energy by electromagnetic waves; also particles e.g. alpha, beta emitted by a radioactive substance.

Reactivity series – an order of reactivity of metals with metal of highest reactivity placed at the top.

Real image – an image that can be projected on to a screen.

Receptor – a sense organ that receives information from the environment.

Recessive (allele) – the allele that is expressed only when a dominant allele is absent.

Red-shift – light from the distant edges of the Universe is moved (or shifted) towards the red part of the visible spectrum; this shows the Universe is expanding or moving away from the Earth.

Reduction – the process in which oxygen is lost, or in which hydrogen or electrons are gained.

Reflection – the deflection of a ray of light when it hits the boundary between two different surfaces, for example, air and glass, as in a mirror.

Reflex – an automatic response to a stimulus; often to protect the body.

Reflex arc – the pathway of a reflex action.

Refraction – the phenomenon that occurs when a wave passes from one medium into another, causing a change in speed and direction (unless the wave hits the second medium at right angles).

Saturated – hydrocarbon molecules that are completely surrounded with hydrogen and have no further available bond sites.

Secondary waves – transverse waves generated by an earthquake.

Sedimentary rock – rock that is formed by consolidated sediment deposited in layers.

Seismic wave – a wave that travels through or along the surface of the Earth as a result of an earthquake, explosion or volcanic activity.

Seismometer – device that detects the vibrations from earthquakes, explosions or volcanic activity.

SETI – Search for Extraterrestrial Intelligence: a scientific experiment with Internet-connected computers collecting data.

Solar cell – a device that is able to transfer light energy into electrical energy.

Solar System – the Sun and the eight planets that orbit it.

Sonar – echo location system using ultrasound waves.

Steady State Theory – the theory that the Universe has always existed in a steady state; that it had no beginning and will have no end.

Sustainable development – a pattern of use of resources that aims to meet the needs of the population, while preserving the environment so that these needs can be met not only for the present, but also for future generations.

Synapse – the small gap between neurones.

Tectonic plates – very large pieces of the Earth's surface that make up its crust.

Thermal decomposition – the breaking down of a large molecule to form smaller molecules by using heat.

Thermal energy – heat energy.

Transformer – a device that changes the size of an alternating voltage.

Transverse wave – a wave in which the oscillations (vibrations) are at 90° to the direction of energy transfer.

Ultrasound – sound waves with frequencies above the upper limit of human hearing, i.e. above 20 000Hz.

Ultraviolet – a region of the electromagnetic spectrum between X-rays and visible light.

Unsaturated – when a long carbon chain molecule contains double bonds.

Unsaturated hydrocarbons – hydrocarbon molecules that contain one or more carbon–carbon double bond.

Variation – the differences between organisms of the same species.

Virtual image – an image that cannot be projected onto a screen.

Viviparous – an animal in which embryos develop inside the body of the mother.

Voltage – the value of the potential difference between two points, such as the terminals of a cell.

Volt (V) – the unit of potential difference or voltage.

Watt / kilowatt (W/kW) – the unit of power, equals the rate of transfer of 1J of energy per second; 1kilowatt (kW) = 1000W.

Wave – a disturbance in a medium or in space that is able to carry energy.

Wave speed – found by multiplying the frequency (Hz) by the wavelength (m) or by dividing the distance travelled (m) by the time taken (s); measured in m/s.

Wavelength – the distance between two successive points on a wave that are at the same stage of oscillation, for example, the distance between two successive peaks.

X-rays – a region of the electromagnetic spectrum between gamma rays and ultraviolet rays; X-rays can be emitted when a solid target is bombarded with electrons.

HT **Black hole** – a body in the Universe with such a large gravitational strength that even light cannot escape; formed at the end of the life cycle of a massive star.

Carrier – an individual who carries a 'faulty' allele, but does not suffer from the condition caused by the faulty allele.

Negative feedback – when a substance is produced that opposes a change to a system, which triggers the return of the system to normal. Hormones in the menstrual cycle and kidneys are examples of this.

Nitinol – a shape memory alloy of nickel and titanium.

Nitrogen cycle – the process by which nitrogen is recycled between living organisms and the environment.

Glossary

HT **Speciation** – the process of evolution by which new species are formed.

Supernova – the release of massive amounts of energy dust and gas into space when a red super giant star explodes.

Vasoconstriction – reducing the flow of blood to the surface of the skin to prevent heat loss.

Vasodilation – increasing the flow of blood to the surface of the skin to encourage heat loss.

Key

	relative atomic mass
	atomic symbol
	name
	atomic (proton) number

1	hydrogen
H	
1	

Group 1	2												3	4	5	6	7	0
																		4 **He** helium 2
7 **Li** lithium 3	9 **Be** beryllium 4												11 **B** boron 5	12 **C** carbon 6	14 **N** nitrogen 7	16 **O** oxygen 8	19 **F** fluorine 9	20 **Ne** neon 10
23 **Na** sodium 11	24 **Mg** magnesium 12												27 **Al** aluminium 13	28 **Si** silicon 14	31 **P** phosphorus 15	32 **S** sulfur 16	35.5 **Cl** chlorine 17	40 **Ar** argon 18
39 **K** potassium 19	40 **Ca** calcium 20	45 **Sc** scandium 21	48 **Ti** titanium 22	51 **V** vanadium 23	52 **Cr** chromium 24	55 **Mn** manganese 25	56 **Fe** iron 26	59 **Co** cobalt 27	59 **Ni** nickel 28	63.5 **Cu** copper 29	65 **Zn** zinc 30	70 **Ga** gallium 31	73 **Ge** germanium 32	75 **As** arsenic 33	79 **Se** selenium 34	80 **Br** bromine 35	84 **Kr** krypton 36
85 **Rb** rubidium 37	88 **Sr** strontium 38	89 **Y** yttrium 39	91 **Zr** zirconium 40	93 **Nb** niobium 41	96 **Mo** molybdenum 42	[98] **Tc** technetium 43	101 **Ru** ruthenium 44	103 **Rh** rhodium 45	106 **Pd** palladium 46	108 **Ag** silver 47	112 **Cd** cadmium 48	115 **In** indium 49	119 **Sn** tin 50	122 **Sb** antimony 51	128 **Te** tellurium 52	127 **I** iodine 53	131 **Xe** xenon 54
133 **Cs** caesium 55	137 **Ba** barium 56	139 **La*** lanthanum 57	178 **Hf** hafnium 72	181 **Ta** tantalum 73	184 **W** tungsten 74	186 **Re** rhenium 75	190 **Os** osmium 76	192 **Ir** iridium 77	195 **Pt** platinum 78	197 **Au** gold 79	201 **Hg** mercury 80	204 **Tl** thallium 81	207 **Pb** lead 82	209 **Bi** bismuth 83	[209] **Po** polonium 84	[210] **At** astatine 85	[222] **Rn** radon 86
[223] **Fr** francium 87	[226] **Ra** radium 88	[227] **Ac*** actinium 89	[261] **Rf** rutherfordium 104	[262] **Db** dubnium 105	[266] **Sg** seaborgium 106	[264] **Bh** bohrium 107	[277] **Hs** hassium 108	[268] **Mt** meitnerium 109	[271] **Ds** darmstadtium 110	[272] **Rg** roentgenium 111							

Elements with atomic numbers 112–116 have been reported but not fully authenticated

*The lanthanoids (atomic numbers 58–71) and the actinoids (atomic numbers 90–103) have been omitted.

The relative atomic masses of copper and chlorine have not been rounded to the nearest whole number.

Notes

Index